A BUZZ
IN THE MEADOW

ALSO BY DAVE GOULSON

A Sting in the Tale

A BUZZ
IN THE
MEADOW

The Natural History
of a French Farm

Dave Goulson

PICADOR   NEW YORK

www.picadorusa.com
www.twitter.com/picadorusa • www.facebook.com/picadorusa
picadorbookroom.tumblr.com

Picador® is a U.S. registered trademark and is used by
St. Martin's Press under license from Pan Books Limited.

For book club information, please visit www.facebook.com/picadorbookclub
or e-mail marketing@picadorusa.com.

Lines from "The Fly" by Ogden Nash, copyright © 1942 by Ogden Nash.
Reprinted by permission of Curtis Brown Ltd.

Extract from West with the Wind by Beryl Markham, reprinted by permission
of Pollinger Limited on behalf of the Estate of Beryl Markham.

The Library of Congress Cataloging-in-Publication Data is available upon request.

ISBN 978-1-250-06588-9 (hardcover)
ISBN 978-1-250-06589-6 (e-book)

Picador books may be purchased for educational, business, or promotional use.
For information on bulk purchases, please contact the Macmillan Corporate and
Premium Sales Department at 1-800-221-7945, extension 5442,
or write to specialmarkets@macmillan.com.

Originally published in Great Britain by Jonathan Cape,
a division of Random House Group Limited

First U.S. Edition: May 2015

10 9 8 7 6 5 4 3 2 1

For Lara

Contents

Preface

In 2003 I bought a derelict farm deep in the heart of rural France, together with thirteen hectares of surrounding meadow. My aim was to create a wildlife sanctuary, a place where butterflies, dragonflies, voles and newts could thrive, free from the pressures of modern agriculture. In particular I was keen to create a place for my beloved bumblebees, creatures I have spent the last twenty years studying and attempting to conserve. This book, in part, is the story of this little corner of the French countryside, of the plants and animals that live there, of their natural history, and of my efforts to encourage them. Most natural-history documentaries and much conservation effort focus on large, charismatic animals: whales, pandas, tigers, and so on. One of my aims in writing this book is to inspire an appreciation for the smaller, everyday creatures that live all around us – the insects and their kin. As chance would have it, many of the insects and flowers that have colonised the farm are species that I myself have studied over the years in my scientific career, and I explain some of the research that has been carried out to explore their secret lives. You will learn how a deathwatch beetle finds its mate; about the importance of flies; how some flowers act as thermal blankets for bees; and about the complex politics of life as a paper wasp, amongst much else. In telling these stories perhaps I can also convey to you the fun of discovery, the satisfaction to be had in teasing apart the details of the lives of the creatures with which we share our planet. More importantly, I want you to realise that what we know and understand about natural history is just the tip of the iceberg. Even among the creatures that inhabit this single meadow, there is no doubt a near-infinite number of beguiling mysteries that have yet to be explained, animals that have never been studied, behaviours that have not yet been observed. What wonders have still to be discovered?

In the second part of the book I show you how the lives of the creatures in the meadow are interwoven with each other and with the wild flowers. Plants compete for space, water and light, are food for herbivores, hosts to parasites and diseases. They use diverse strategies to tempt pollinators to visit them, and in turn their pollinators have evolved numerous tricks so that they can learn which flowers are most rewarding and can gather those rewards quickly, sometimes robbing their hosts, at other times being duped into pollinating flowers for no reward. Plants depend on a horde of small animals and microorganisms to break down leaves and dung to release their nutrients, and they benefit from the actions of predatory birds, spiders and insects that keep down the numbers of caterpillars, grasshoppers and greenfly that might eat their leaves. Every species is linked, one way or another, to hundreds of others, in a web of interactions that are at present far beyond our ability to fully comprehend.

In the final part, I explain how the modern world has become increasingly inhospitable for wildlife, as humans squeeze ever more from the land to provide for our many needs. I give some examples of the devastation we have caused – and are causing – to our planet, from the effects of primitive man's prehistoric spread out of Africa to the insidious damage that we continue to do through our overuse of poisonous chemicals in the countryside. Many of the fascinating creatures with which we share our world are slowly disappearing as a result of our actions, often before we have learned a single thing about their lives, or of their role in the tapestry of life. This book is intended as a wake-up call, to remind us that we should cherish life on Earth in all its forms. As species become extinct, so the mysteries of their lives are lost for ever. We are destroying our children's inheritance, stealing from them the joy of discovery and exploration of the natural world. What is more, we are undermining the ability of our planet to support us; although we understand very little about the myriad complex interactions between the many creatures on this Earth, we do have good evidence to suggest that these interactions are vital to the health of the planet, and hence are vital for our own well-being and perhaps for our very survival.

I want to make you look at our world with new eyes; to persuade you to go out into your garden or a local park and get down on your hands and knees and *look*. There is so much to see. If you look closely enough, you cannot fail to begin to appreciate the precious undiscovered glory that is life on planet Earth. If we learn to value what we have, then perhaps we will find a way to keep it.

A BUZZ
IN THE MEADOW

PART I

Tales from the Meadow

We inhabit a spherical rock, just 13,000 kilometres miles across, floating in the unimaginable vastness of space. It is at least ten thousand billion kilometres to the nearest planet that might possibly support any other life, a distance of which our brains cannot begin to conceive. We spend much time and effort on building telescopes that can look ever further into the void, and on listening to and analysing radio waves from distant galaxies, in the hope of detecting signs of other life forms. Many films, TV shows and novels speculate about what might be out there. Yet there are real wonders of the universe right here, all around us, and we pay them little heed. We are lucky enough to share our little rock with perhaps ten million different species, and many of them have not yet even been given a name.

I am fortunate enough to own a small hay meadow in rural France. Being something akin to the entomological equivalent of a train-spotter, I have so far identified more than seventy bee species, fifty types of butterfly, sixty bird species and well over 100 different flowering plants living in this meadow. This is just a small fraction of the grand total; I have not yet begun to tackle the springtails, mites, worms, spiders, beetles, snails and other creatures that live there, and in all likelihood I will never find time. The vast majority of the creatures that we ignore are small, many so diminutive that they can barely be seen with the naked eye, and others much smaller still. But if you take the trouble to place one of these minute creatures under a microscope you will reveal their precise symmetry and exquisite structure. Each and every one has a different story, a life history; it must find food, grow, evade predators, find and court a mate, lay eggs, and so on. Every step involves challenges, obstacles that must be overcome, and every species has evolved its own unique combinations of strategies to survive and thrive; if it had not, it would long since have disappeared. Even in western Europe, where we have a long tradition of studying natural history, we know almost nothing about the lives of most of these wild creatures.

In this section I will introduce you to some of the insects and other small animals that live in this meadow, to some of the very few that have been studied at least a little, and to what is known about some of their relatives that live in more exotic climes. I will try to explain some of the fascinating details of their behaviour and ecology, what roles they play in the ecosystem, and my own efforts to encourage more and more species to colonise this little corner of the French countryside. Welcome to the meadow . . .

CHAPTER ONE

A Stroll in the Meadow

24 April 2007. Morning run 5.8 miles, 42 mins 2 secs. As ever, the French countryside was almost devoid of human life; I saw no people, but was barked at by five dogs, unused to seeing a runner passing by. It was a lovely cool morning, clear blue sky above, thick dew on the grass, cowslips bursting from the hedge banks. Butterfly species seen: 6 – I distract myself from the pain of running by seeing how many I can spot without stopping. I've tried this with bumblebees, but they are mostly too tricky to identify at speed. Today's butterfly haul included a holly blue and a male brimstone, sulphurous wings flashing in the sunshine. I also disturbed a pair of green woodpeckers anting on the lane above the top field, their alarmed yaffle and undulating flight unmistakable. Lesser whitethroats were singing in every copse I passed, a melodic, liquid song; the mating season is clearly in full swing – I can still hear them from all directions as I sit on the patio bench by the front door, dripping sweat on to my notes.

Sixty-five kilometres north-west of Limoges, near the lovely Roman market town of Confolens on the River Vienne, stands an old farmhouse. Roughly halfway down France, going north to south, and about 110 kilometres inland from the west coast, the farmhouse lies in the Charente, a large, sleepy *département* of rolling countryside, oak forests, rust-coloured Limousin cows, and fields of sunflowers, intersected by the lazy meanders of the Charente River. The house was built perhaps 160 years ago, presumably by a Monsieur Nauche who gave the farm its name, Chez Nauche. There are many grand and beautiful Charentais farmhouses in the region, built of dressed stone three or more storeys high, with ranks of tall windows arranged symmetrically around an imposing central entrance. This is not one of them. At Chez Nauche the thick walls are built from undressed, local limestone, irregular lumps of rock full of fossils and presumably dug from the local fields. The stones are held together with orange clay for mortar, also dug straight from the ground. The walls have shifted since they were built, and now lean at interesting angles. The windows are mostly small and irregularly arranged, with ancient weathered oak beams for lintels and loosely hinged old oak shutters from which the paint has largely peeled. The house is long, low and squat, facing south; the intention was that all accommodation should be on the ground floor, a common design among the more modest farmhouses in the area. The large attic was for hay storage, which provided insulation during the winter for those living below. The floors to the attic are made from thick planks of oak, laid upon massive square oak beams. The timber would mostly have come from local trees, hand-sawn, and indeed the beams still bear the saw-marks. The labour involved in building a house like this must have been Herculean, although the costs of material would have been close to zero.

To produce an oak beam, the practice was simply to find the nearest oak tree with a fairly straight trunk and chop it down. The builders would then dig a pit under the fallen trunk, deep enough for one of them to lie in, and they would saw the trunk into square beams using a huge two-man saw, with one person lying in the pit, his face sprinkled with sawdust, and the other standing on top of the trunk. Finally they would use a horse to drag the beam to the house, and ropes to winch it into position.

The terracotta tiles on the roof are also fired from local clay. They are known as canal or channel tiles, a design that dates back to the Romans, and are laid in alternating rows of gulley and ridge. I doubt that Monsieur Nauche made those himself, since firing them is a bit of a specialist job, so they are probably one of the only major items that he had to buy in, but they would not have come from far away. Otherwise, pretty much the entire building, and its surrounding barns, was constructed from materials that could be gathered for free from the immediate surroundings, and this gives the buildings a natural, organic feel, almost as if they grew up from the ground of their own accord like an eruption of unusual, rectangular mushrooms.

I bought Chez Nauche in 2003, from an old farmer named Monsieur Poupard. So far as I could establish with my feeble grasp of French, he had lived there all his life, keeping dairy cows and growing arable crops. Well into his sixties and with no children to leave the farm to, he had decided to sell up and retire. He had not looked after the old place, allowing it to fall gently into ruin. The roof leaked, so that the internal timbers were slowly rotting, and the old lime plaster was stained black with mould and was peeling from the walls. The window frames were rotten, the glass was cracked and covered with patches of old plastic sheeting, and the front door was rotted away at the base, with old pieces of tin can hammered flat and nailed over the gaps. The plumbing consisted of one old dripping tap above a stone sink – there was no bath, shower or toilet, and the lavatory facilities consisted of a bucket in the shed.

It was, to put it mildly, a doer-upper, but for all its shortcomings it held one huge attraction for me, as a wildlife-obsessed biologist. Monsieur Poupard's lackadaisical maintenance schedule had allowed the house and its surroundings to be infiltrated by a myriad of creatures. In many modern British houses, house-proud home-owners are horrified if they see a single woodlouse on the carpet, or an ant in the kitchen. This attitude must swiftly be abandoned at Chez Nauche, or a nervous breakdown would inevitably ensue. The house has slowly settled into its environment over the decades, and is swamped and overrun with plants and animals. Although I have made some improvements in the ten years since I bought it, it remains to this day a haven for wildlife. The roof tiles are crusted with orange, black and cream lichens, which are grazed upon by caterpillars. Mosses grow in the gullies between the tiles, particularly on the north side of the house, and millipedes, woodlice, water bears* and numerous other small insects live amongst the damp green cushions. The walls are also encrusted with lichens, and are smothered under the lush foliage of the grape vines that cling to rusting metal brackets along the wall. When the sun shines, as it often does, these walls are a popular basking spot for butterflies, bees and flies, warming themselves before going off to look for a mate or nectar to drink. These insects are hunted by zebra-striped jumping spiders and mottled brown-and-green wall lizards, agile creatures

---

* It is quite likely that you have never heard of water bears, also known as moss piglets or, more properly, as tardigrades. These tiny, eight-legged creatures, which rarely exceed one millimetre in length, are amongst the hardiest animals on Earth. They can survive a decade without water, being cooled to -273°C, heated to 150°C, crushed at 6,000 atmospheres pressure or exposed to 1,000 times more radiation than would kill a human. I have absolutely no idea why scientists have taken it upon themselves to try so hard to kill these innocuous little creatures.

with long, clawed toes that scurry impossibly quickly over the vertical masonry, dashing into holes in the soft clay mortar at the first sign of danger. Most of the insects are too quick to be caught, especially if they have managed to keep warm and ready for take-off, but once in the air they run the gauntlet of the swallows that nest in the barns and swoop low past the house. From the base of the wall at the front of the house sprout old lavender bushes, their twisted, woody stems sagging under the weight of purple blossom in summer, alive with bumblebees, butterflies and the blurred wings of hovering hummingbird hawk-moths, their long crooked tongues reaching down into the nectaries of the flowers.

An old cobbled path runs to the front door, and the cracks between the stones are inhabited by bulbous-headed black crickets, the males singing cheerfully and incessantly to attract a mate. The lizards and young western whip snakes also make use of the holes amongst the warm stones, hunting there for beetles and spiders. In front of the house is a stooped and gnarled selection of ancient nectarine and plum trees, with bracket fungi sprouting from some branches, and chubby green caterpillars of the scarce swallowtail grazing on their leaves. Great green bush crickets perch on the branches, the males rasping out their incessant chainsaw-buzz in an attempt to drown out the black crickets down below.

Inside the house, where it is cool and dark and the buzz of the crickets is just a distant hum, crepuscular creatures abound. Spiders of numerous species spin their webs amongst the ancient beams; spindly daddy-long-legs spiders spin irregular, shoddy webs from which they dangle upside-down, while giant *Tegenaria* house spiders prefer to make close-woven, funnel-shaped webs leading to a deep hole in which they can hide. The beams themselves are tunnelled by the fat white grubs of long-horn and death-watch beetles, and also by woodworm (not a worm, but a tiny beetle). Under the furniture and in the kitchen cupboards lurk satin-black darkling beetles, ponderously slow but heavily armoured, so they have no need for speed.

At night, the mice take over; on the floor, house mice scurry, with the occasional larger, huge-eyed wood mouse. They search for scraps of human food, tasty spiders or day-flying insects that have blundered into the house and become trapped. On the walls and beams, dormice scamper: garden dormice, with delicate racoon-like facial markings and a long tail ending in a fluffy tip; and the scarcer edible dormice, favoured as a delicacy by the Romans. Endearing to look at they may be, but the garden dormice are aggressive little beasts, churring at each other through the night, and they often wake me with their rumbustious skirmishes. Because of the nuisance they make of themselves, I have trapped many dozens of them; they are absolute suckers for Cantal, a hard and pungent cheese from the mountains of the Auvergne – it gets them every time. When my eldest boys Finn and Jedd – at the time about seven and five years old – first saw one of these garden dormice, growling angrily at them from the trap and gnawing at the mesh to escape, they rushed to wake me up with the news: 'Daddy, come quick, we've caught a tiny demon!' It did look pretty ferocious – the poor thing had rubbed its nose red-raw trying to get out. I always release the little demons far away from the house, having given them a good feed, but my efforts never seem to make any dent in the population. The edible dormice seem to be much gentler, with a beautifully thick fluffy tail; they are so large as to be easily mistaken for small, exceedingly cute squirrels. I cannot bring myself to evict them from the house.

The various mice are nervous, for barn owls roost in the attic, leaving huge piles of pellets, which are consumed by the grubs of clothes and skin moths, species adapted to feeding on the desiccated remains of animals. There is also another, mysterious beast that they should fear. Some years ago I installed some Velux windows in the old roof, and soon afterwards noted the footprints of a largish animal on the glass. I also found pungent, elongated scats, sometimes on the drive to the house, and once on an inside windowsill. Whatever this beast was, it could take on formidable prey; on one occasion I found a wing and the head of one of my barn owls strewn in the attic. On another occasion, when on an early-morning excursion, my young boys found a bleeding chunk of flesh on the drive, all that remained of a large whip snake. From its width I would guess the snake had been a good one and a half metres or more long, but everything had been consumed, apart from a fifteen-centimetre section of its midriff. The beast took on a mythical status in the family, with the children speculating wildly as to what it might be, and it was many years before I finally worked out what it was.

Let me take you for a stroll. We'll start at the top of the drive, to the north of the house, by the big horse-chestnut tree. It is late afternoon, towards the end of May, and the tree is in full bloom, the cones of frothy cream flowers attracting scores of bumblebees, whose bustling dislodges petals from the older flowers that rain down upon the drive. We amble down the old tarmac drive, its warm surface cracked by tree roots pushing through from beneath, sparse tufts of crested dogstail grass sprouting from the crevices. On the left we stop to admire the wood-ant nest, a gentle dome of cut, dried grass stems thronging with large chestnut-coloured ants. The nest has been in the same place for ten years now, to my knowledge. My boys love to watch and poke the ants, and occasionally, I suspect, they throw them insect prey. The slightest disturbance causes ripples of activity to spread across the nest as the ants release alarm pheromones warning of danger. The ant trails radiate from the nest across the tarmac, with incoming ants carrying all sorts of fragments of plants and insects to feed to their brood in the nest.

Beyond the ants' nest on our left is a thick hedge of gorse, five metres or more across. A male stonechat perches on the highest point, his trademark call sounding very much like two dry pebbles being struck together. The female is no doubt sitting on her cup-shaped mossy nest somewhere deep in the gorse thicket, incubating her clutch of sky-blue eggs. Peering through the thick gorse hedge, to the east of the drive we can just see my orchard: fifty well-spaced young apple trees that I grew from pips. The largest are now nearly four metres tall, and two of the trees bore fruit for the first time last year. My three boys are chasing butterflies fifty metres away amongst the trees, the two eldest, Finn and Jedd (now aged twelve and ten) leading the way through the long grass, chattering excitedly, each armed with a huge kite net. Behind them our youngest, Seth (aged three), is gamely battling to keep up, his white-blond shock of hair all that is visible of him amongst the greenery.

On our right I point out a bee orchid, its single purple flower mimicking the smell and texture of a female bee and thus luring male bees to attempt to copulate with it. All they get for their trouble is a ball of pollen glued to their heads, but they must be foolish enough to make the same mistake again or the bee orchid's strategy would not work.

Further down, the drive is shaded by a line of large oaks on the right, and a mix of elm and oak on the left. Brittle brown acorns from last autumn still litter the ground. The elms are repeatedly attacked by Dutch elm disease, which quickly kills the trees once they reach six or seven metres in height, but luckily the trees spread rapidly by suckers, so there is a constant crop of new saplings coming up. A territorial male speckled wood butterfly dashes up from a warm sunspot on the drive to chase away a brimstone that has dared to enter its domain.

I love the French names for butterflies, compared to which many of the English names are a little unimaginative; for example the English *orange tip* is simply descriptive, while the French *l'aurore* – the dawn – is rather more poetic. What do we call a speckled butterfly that lives in woods? The *speckled wood*, of course, while to the French it is *le Tircis*, named after a shepherd in a seventeenth-century fable by Jean de La Fontaine. A few years ago I hit upon the idea of organising a guided butterfly walk at Chez Nauche for any interested locals. I sent posters advertising the walk to the mayor of Épenède, the local village, and also to the mayor of nearby Pleuville, asking for them to be displayed on the village noticeboard. I bought lots of lemonade for my visitors, and boned up on all the French names of butter-flies and other insects, although I was somewhat worried that my inability to say much else in coherent French might be a handicap. On the day of the event I waited nervously outside the house, but no one arrived at the allotted time. Ten minutes late a car at last drew up; an English lady, and her young daughter, who lived nearby. I had not met them before, but was happy to take them for a walk in the meadow, though also a little disappointed by the turnout of the French contingent. Perhaps chasing butterflies is an eccentric English activity, and not something that appeals to French country-dwellers. It is certainly true that membership of conservation charities such as the RSPB and Butterfly Conservation is far higher in the UK than in any other country in the world. We had a pleasant walk, spotting bumble-bees, butterflies and grasshoppers. Towards the end of the walk I took us past an old piece of corrugated tin that I had laid out on the edge of the field. Snakes love to bask under tin sheeting, and I had a pretty good idea that there would be something dramatic underneath, to form the perfect finale to the walk. Sure enough, there was a sizeable Aesculapian snake underneath, which I managed to grab with a flourish. We walked back to the car so that the mother could take a photo of her daughter stroking the snake, and finally we let it go. I hadn't quite anticipated what happened next. The snake shot under their car, then climbed up into the still-warm engine. We spent the next hour with the bonnet up, trying to find it – without success. In the end the poor lady and her daughter had to drive away reluctantly with a snake somewhere in their car. I very much hope they all survived the journey.

Returning to our stroll, we are coming towards the end of the drive. On our left is a rectangle of stout walls – the Alamo, as my father has christened it – all that remains of a very large barn. When I bought Chez Nauche this barn was in a terrible state, with gaping holes in the roof and the beautiful old oak frames well rotted. I couldn't afford to repair it, so I took the roof off and sold the remaining half-decent timbers to a reclamation yard. The old walls provide a suntrap for lizards and warmth-loving butterflies; teasels and thistles sprout up in profusion from the stony ground; and whip snakes are common amongst the stones and weeds.

On our right is a small hollow, overgrown with blackthorn and ash, once a shallow seasonal pond, which I mistakenly filled in with building rubble. I have since been slowly clearing it out, in the hope that the newts that once lived there will return.

Let us strike right off the drive, past the pond and across the open meadow. This western side of the meadow is where I have set up a large, long-running experiment to try to increase the numbers of flowers. I sowed squares of meadow with yellow rattle, eyebright, bartsia and meadow cow-wheat, all partially parasitic plants that sap the strength from nearby grasses by tapping into their roots and sucking up nutrients. Suppressing the grasses leaves a little more room for other flowers, or so the theory goes. The rattle is in full flower: a pretty annual with small yellow flowers tipped in purple, which has established itself in little clumps across the experimental plots. It is too early to say whether this has increased the number of flowers, but in any case the meadow looks pretty good at this time of year. After ten years without any fertilisers or pesticides, quite a lot of wild flowers have established themselves. The main grasses are cocksfoot, Yorkshire fog and false oat grass, large and dominant species that tend to smother all else, but over time they have been declining and have been partly replaced by the finer, less aggressive grasses typical of a proper hay meadow: fescues, sweet vernal grass and meadow foxtail. Amongst the grasses, some flowers have become common: wild geraniums, forget-me-nots, ragwort, white campions, hawkbit, clover and meddicks, to name but a few. Some of them tend to occur in distinct patches, either because their seeds do not spread readily or perhaps because some subtle variations in the soil properties suit them better in some places than others.

As soon as we leave the drive we enter a patch thick with cinquefoil, a low-growing, prostrate relative of the rose, with simple yellow flowers, much like those a child might draw. Its creeping, horizontal stems snag our feet as we walk through. Five metres later the cinquefoil ends abruptly, and we encounter a dense clump of meadow vetchling, a pea with twining tendrils with which it clambers up the taller grass stems. Amongst the close vegetation we hear the high-pitched shrieks of shrews fighting; these tiny but voracious predators live their short lives at a hectic pace, eating constantly and fiercely defending their territory against one another. After the vetchling, a dense patch of red clover is thick with long-tongued bumblebees, garden bumblebees and common carder bumblebees, gathering its protein-rich, toffee-coloured pollen and sweet nectar. Then we move into a dense sward of lady's bedstraw, a fragrant spreading plant with tiny, dark-green leaves and heads of abundant but minuscule yellow flowers. In days gone by, before comfy sprung mattresses, it was used as sweet-smelling bedding – whence, of course, it gains its name.

We are walking south-west, down a gently increasing slope, with the old farm buildings of the tiny hamlet of Villemiers visible on the other side of the valley a kilometre away. The Transon meanders in the bottom of the valley below, a lazy trickle of a stream with small muddy pools at intervals, home to numerous coypu, a South American rodent that escaped from fur-farms long ago and has found a home-from-home in the many rivers and lakes of the Charente. They are semi-aquatic, resembling beavers in all but their long, rat-like tails. They can be something of a nuisance, as they are great burrowers, creating huge holes in the banks just on the waterline, which does little harm in a stream, but can be disastrous in a man-made lake, since their burrows can puncture the dam.*

Away to our left, the plaintive cry of the wack-wack bird can be heard in the distance. My boys and I have spent many hours trying to stalk this beast, which I have only ever heard at Chez Nauche. It calls most days in spring and summer, usually from the south-east, a nasal wack, wack with a distinct but brief pause

---

* The UK also used to have escapee coypu in East Anglia, accidentally introduced in the 1920s. They caused havoc by burrowing through the banks of the many drainage ditches and canals in this very flat part of the UK, often causing fields to flood. In 1989 I met a scientist who had recently taken a job at the MAFF-funded coypu control centre at a time when, although no one had yet realised it, coypu had already been successfully exterminated; the last one was seen in Norfolk in 1988.

between the notes. There only ever seems to be one of them. Whenever I try to do an impression of it to my knowledgeable ornithological friends, they laugh and tell me it is a duck, but that is simply my inability to replicate the noise. We have crept towards the source of the noise through the long grass of the meadow. It usually sounds as if it is coming from a large oak tree on the boundary, but whenever we get close it ceases to call, and we see nothing fly away. The boys speculate that it is some dramatic creature, brightly coloured and a metre or so tall, with a crest and a long sharp beak, but if so, it must be very good at hiding. I wonder whether it may not be a bird at all, but some peculiar species of frog. Perhaps one day we will find out.

The meadow becomes drier as we continue on to the steep south-facing slope at the southern end, and ribwort plantain becomes common underfoot. This is an unspectacular little plant, with strapline leaves and inconspicuous brown flowers from which dangles a fringe of yellow anthers, but the leaves are the favoured food plant of the lovely Glanville fritillary. This butterfly is named after Lady Eleanor Glanville, one of the very few female lepidopterists of the eighteenth century. She first described this pretty species, which she found near her home in Lincolnshire. Glanville fritillaries have long since disappeared from most of the UK; they are now found only on the south coast of the Isle of Wight, but it is one of the most common butterflies at this time of year at Chez Nauche, and we disturb dozens from the grass as we walk. They have an orange-and-black chequerboard upper side to their wings, their creamy underside being streaked attractively with orange and dotted with black spots. Their bodies are furry, giving them a rather cuddly appearance. I bred Glanville fritillaries in my bedroom as a child, after buying the pupae from Worldwide Butterflies, and I have always been rather attached to this species.

The caterpillars are unusual in that they are gregarious; the female lays large mounds of yellow eggs, which hatch into velvet-black caterpillars, which live together on plantain in silken webs that they spin. Once they have consumed the plant on which they are laid, they somehow agree that it is time to depart and set off in a convoy to the next one.

We are approaching a deep-sunk green lane that marks the western boundary of the meadow. A dense stand of oak, hazel and blackthorn lines both sides of the lane. We push through a slight gap in the hedge, our legs getting scratched by the terrifically spiky butcher's broom that thrives on the hedge bank. In the lane it is shady and sheltered; on hot days flies congregate here to escape the heat. I have brought us through to see the wood whites, delicate, ghostly-white butterflies that patrol slowly up and down the lane, their flight so weak it seems they may expire at any moment. This is another species that is in precipitous decline in the UK for reasons that are not well understood, but here they seem to be flourishing. We turn left down the lane, continuing steeply downhill to the Transon, a stream that is just beyond my land. There is a small pool before it gurgles under the lane, and a swarm of shiny whirligig beetles gyrates crazily on the surface. I've often seen grass snakes hunting fish and tadpoles in the shallows here, but there isn't one today. Just as we turn to retrace our steps a male demoiselle flits by, its metallic blue body glinting in the sunlight. This is the king of damselflies, larger than other European species and by some margin the most spectacular. Aside from the male's iridescent body, its wings are decorated with large splashes of blue-black pigment, so that they flash with every wingbeat. The females are a slightly more understated iridescent green, and a pair sitting together, as they often do, is a breathtaking sight.

We walk a little way back up the hill and cut back through

the hedge into the south corner of my meadow. We climb up a steep slope, heading north-east, towards a small tree standing in isolation. It is a walnut that I planted there some six years ago, now grown to about three metres in height. One day it will be large enough to make a splendid shady picnic spot, and perhaps also provide walnuts to eat. On the slender grey trunk there is a praying mantis, newly adult, its triangular head following our every movement, as if sizing us up as potential prey. In green vegetation praying mantises are nearly impossible to spot, but this one has chosen the wrong place to perch. Its powerful forelegs are folded beneath it, their rows of sharp spines locked together, poised to strike out in the blink of an eye, should an insect be foolish enough to come too close. If attacked by a bird, the mantis can flash its wings open, revealing large eye-spots, designed to frighten into retreat all but the boldest bird.*

Just beyond the walnut is a gentle hollow perhaps twenty metres across. Here the grass is thick with wild basil, thyme and mint, which create a heady aroma. Sitting down, you cannot be seen from anywhere; it is a wonderful place to relax and soak up the sights, smells and sounds of the meadow. A male stag beetle drones past; they are common at this time of year. These huge beetles are clumsy fliers, encumbered as they are with massive jaws for

---

* Due perhaps to their large size and striking appearance, mantises have long been associated with all sorts of odd beliefs. In North America it was commonly held that they could blind men and kill horses. The French regarded them as more benign, believing that they would point the way home to lost children, while in parts of Africa they are thought to bring good luck and occasionally resurrect the dead. Not bad for an insect that is but one step removed from a cockroach (they are close relatives).

wrestling with rivals for a mate. They are so slow that it is easy to snatch them out of the air, but I leave this one be.

From here we head east, the meadow falling away again into a gentle valley, at the bottom of which is a small spring. The spring was once the main water supply to the farm. French water is metered and amongst the most expensive in the world, so Monsieur Poupard used to pump all of his water up from the spring to a rusty old tank in one of the small barns, thereby avoiding having to pay for it. A well has been dug into the ground, lined with stones, and from this a trickle of water runs south towards the Transon. I have allowed scrub to establish around the spring, mainly blackthorn and brambles, which provide a glorious impene-trable tangle in which many birds nest. Slightly downstream I planted yellow flag irises, which have taken well and sprout their waxy leaves and stems well above the encroaching brambles, their flamboyant flowers a draw to bumblebees.

Beyond the irises we come to a pond held back by a clumsy stone-and-clay dam, my attempt to create more habitat for aquatic wildlife, of which I will tell you more later. We walk across the top of the dam and up the other side of the valley, still heading east. On our right, the boundary is marked by huge mature oaks, alive with the bubbling, liquid song of whitethroats. When we reach the top of the hill we are near my eastern boundary. We sit down, looking back over the valley and the spring, to the cluster of ochre buildings that make up Chez Nauche, casting long shadows towards us as the sun falls behind them to the western horizon. A swallowtail butterfly soars past, the first of the year, a magnificent yellow-and-black creature, the hindwings of which are decorated with blue-and-red eye-spots and long streamers. It is a male, searching eagerly for a newly emerging female with which to mate. The crickets, which fell silent as we approached, edge back to the mouths of their burrows and recommence their singing. Summer is near, and for insects this is the time for sex and nectar, sunshine and flowers. It is my favourite time of year, and my favourite place, where nature runs riot and all is right with the world. Well, almost. If only I'd remembered to bring a couple of cold beers. And perhaps a nip of cheese.

CHAPTER TWO

The Insect Empire

27 July 2007. Run: 41 mins 15 secs. It is another beautiful day in paradise. People: one old man delivering bread from his white Citroën van in Épenède. Dogs: 8 – a personal record, including a huge Pyrenean mountain dog in Le Breuil, with a bark that made the earth shake. Fortunately it seemed friendly. Butterfly species: 16. Black-veined white butterflies are plentiful this year, braving the spiky flowers of teasels along the drive to gorge on the rich nectar; for mysterious reasons, this butterfly species died out in the UK more than 100 years ago. As I sit, getting my breath back, I can see a Montagu's harrier hawking above the newly cut top meadow, hunting for voles – a magnificent, graceful, but angular bird with slate-grey, black-tipped wings.

> *We hope that, when the insects take over the world, they will remember with gratitude how we took them along on all our picnics.*
>
> Bill Vaughan

Insects are creatures with three pairs of legs,
Some swim, some fly; they lay millions of eggs.
They don't wear their skeletons in, but out.
Their blood just goes sloshing loosely about;
They come in three parts. Some are bare; some have hair.
Their hearts are in back; they circulate air.

They smell with their feelers and taste with their feet,
And there's scarcely a thing that some insects won't eat:
Flowers and woodwork and books and rugs,
Overcoats, people, and other bugs.
When five billion trillion keep munching each day,
It's a wonder the world isn't nibbled away!

Ethel Jacobson, 'The Insects' World'

Half a billion years ago, give or take, a slow revolution began. On the muddy floor of an ancient ocean, a selection of weird and wonderful creatures began their bid to take over the world. Most of them looked little like any of today's living creatures. They had segmented bodies equipped with a varying array of tentacles, claws, spines, eyes and numerous other odd appendages, the purpose of which we will almost certainly never understand. We would not know about these wonderful, long-dead creatures were it not for the diligence of one Charles Walcott, a fossil-hunter and geologist who, late in his life in 1909, stumbled upon a huge selection of beautifully preserved fossils from this era high in the Canadian Rockies. The rock formation, now known as the Burgess Shale, was formed from layers of soft silt that had settled on the ocean floor, trapping and preserving in extraordinary detail the bodies of the creatures that lived there. We still don't know why the fossils here were preserved so well; it may have been that an area of the ocean floor was anoxic, so that creatures entering it suffocated and were preserved by the lack of oxygen, or it may have been that a series of sudden mud-slides trapped and preserved these hapless creatures. Whatever the reason, the Burgess Shale provides a remarkable picture of a primordial world.

Walcott spent the last years of his life in repeated trips to the Burgess Shale, and in attempting to identify and classify the fossils he collected. Most of the primitive animals that he described he assigned to one group, the arthropods (meaning 'jointed feet'). This is the group that today comprises crustaceans, arachnids and the insects. All arthropods have a segmented exoskeleton, a rigid, articulated suit of armour, usually equipped with an array of jointed limbs. They have been compared to Swiss-army knives; their limbs can each be specialised for different functions: walking, swimming, grabbing, stabbing, mating, breathing, flying, weaving and so on. Just like the army knife, these limbs often fold neatly away when not in use.

Among the arthropods in the Burgess Shale were crustacean-like creatures, relatives of the crabs, lobsters, krill, shrimps, barnacles and copepods that abound in the seas to this day. There were many other creatures too, ones that were hard to classify into familiar arthropod groups and presumably belonged to lineages that did not survive the intervening eons to the present day. There was *Opabinia*, which appeared to have five eyes and a downcurved trunk like that of a minature elephant; and *Hallucigenia*, a surpassingly strange creature a little reminiscent of a cross between a worm and a hedgehog, with numerous legs and paired sharp spines. When this was first described, it was thought to walk on its spines, with its tentacle-like legs waving above it. It has now been flipped over and is portrayed with its spines on top, presumably as a form of defence, although we will never know for sure. One of the larger genera, named *Animalocaris*, was frequently preserved in fragments, perhaps because its body easily broke apart after death, and its different body parts were originally classified as three different animals, until a whole specimen was eventually discovered.*

---

* In the 1970s a PhD student at Cambridge University named Simon Conway-Morris began a re-examination of Walcott's fossils, and concluded that many of them represented forms of life quite distinct

Since the discovery of the Burgess Shale, other similar fossil beds from about the same period have been found elsewhere in the world, although arguably none quite so fine. These have added to our knowledge of life in the Cambrian seas half a billion years ago – 499 million years before something approximating to modern humans was to appear. In addition to Walcott's strange creatures there were arachnids, ancestors of modern spiders and ticks, including the fearsome eurypterids, scorpion-like creatures up to two and a half metres long, which used powerful pincers to hunt their prey on the beds of oceans and rivers. There were trilobites, segmented, shield-shaped animals known to us now only from

---

from any known today. He argued that most were not arthropods, or members of any other surviving animal phylum, but that they belonged to a range of different phyla unknown to science. The prominent and eccentric evolutionary biologist Stephen Jay Gould of the University of Harvard jumped upon this as evidence for his theory of 'punctuated equilibria'. He argued that the evolution of life consisted of long periods when nothing much happened, interspersed with sudden bursts of activity when weird and strange new life forms were thrown up in profusion, as in the Burgess Shale. He contrasted this with the traditional view that evolution was a gradual process. This led to fierce debate, with many biologists pointing out that there was no strong evidence that the fossils in the Burgess Shale – or anywhere else for that matter – had appeared suddenly. The two camps became known as 'evolution by jerks' versus 'evolution by creeps'. With hindsight, it was all a little silly, for no one had ever claimed that evolutionary change should take place at a steady pace, and nor is there any reason to expect this. More recent evaluation suggests that Conway-Morris was mistaken and that most of the animals in the Burgess Shale are indeed arthropods, or their close relatives, however strange they may appear.

fossils, but for 250 million years they were amongst the most abundant creatures on Earth, with many thousands of species, from tiny free-swimming versions thought to have lived in open water to vast, armoured bulldozers that trundled along the ocean floor, presumably trying to keep out of the way of the eurypterids.

One arthropod group that is conspicuously absent from the fossils of the Cambrian is the insects, but that would have been no surprise to Walcott, for the insects did not evolve in the seas; they evolved later, on land. At the time the Burgess Shale was laid down, life on land was pretty unexciting – there were a few primitive clubmosses and liverworts growing in wet areas near water, but little else. There would also have been a strand-line of washed-up plants and animals, edible detritus that was tantalisingly out of reach of the water-bound animals of the day. Inevitably, before long some arthropods began to drag themselves ashore to take advantage of these untapped resources. Their external skel-etons, which perhaps originally evolved as a defence against preda-tors, helped to support their bodies on land, giving them an edge in colonisation there over soft-bodied sea animals such as jellyfish and worms. To start with, these animals would presumably have been poorly waterproofed and had to return to water very regularly to prevent themselves drying out; or they would have had to stay in the dampest places, amongst rotting weed or damp moss.

We will probably never know what the first land animal looked like; it may have been something like a millipede, which grazed on the carpets of moss on the sea shores. It may have been a crustacean, perhaps similar to a woodlouse or sandhopper, feeding on rotting detritus along the high-tide line. Whatever it was, it was soon followed by predatory arachnids, scorpion-like creatures that trundled or scurried after their prey across the greenery. As the plants slowly adapted to life on land and spread away from the seas, they also grew taller as the competition for light intensified. To follow them, the animals had to improve their waterproofing. Some animal groups never really got the hang of this; for example, the few crustaceans that successfully invaded land, such as woodlice, are restricted to damp places to this day. Others, such as arachnids, evolved more-or-less waterproof cuticles and eggs, so that they were able to leave water far behind and occupy even the most inhospitable, arid environments on Earth.

The insects were the last of the major arthropod groups to arrive, about 400 million years ago, but they have more than made up for lost time since then. We do not know what they evolved from – perhaps it was the crustaceans, perhaps some early millipede-like creature. It is most likely that they evolved on land rather than in the sea like the others, and they have become the masters of terrestrial life.

Along with the spiders, insects mastered waterproofing early on; their cuticle is coated in waxes and oils that cut water loss to a minimum. Insects differ from other arthropods in having fewer legs – just six. Their body is divided into three distinct sections: the head, which carries the sensory organs: eyes, antennae, palps, and so on; the thorax, to which all the limbs are attached; and the abdomen, which contains the reproductive parts. The earliest insects were not particularly impressive. They were probably similar to the silverfish that survive to this day: small, scurrying creatures that live in damp places, notably under carpets in poorly maintained houses, presumably not their original habitat. Somewhere along the line these early insects acquired better waterproofing and, with that, the terrestrial world was their oyster. They proliferated, exploiting the abundant food provided by the spreading forests and specialising into a myriad of forms. By the Carboniferous period about 360 million years ago, there were numerous types of cockroach, mantis, grasshopper and probably many others; insects rarely fossilise, so we have only a very fragmented picture. These insects provided abundant food for the predatory and parasitic arachnids, and they too thrived and became better adapted to life on land, so that the Carboniferous was blessed with the first spiders, and also with bloodsucking ticks, scavenging harvestmen and mites, and various other horrendously unattractive but eerily fascinating creatures, such as whip scorpions and vinegaroons, which have survived to the present.

The insects were to prove to have a couple of other tricks up their sleeve – two more evolutionary innovations that would leave the rest of the animal kingdom far behind. Perhaps most importantly, they were the first creatures on Earth to take to the air, to evolve powered flight, perhaps 350 million years ago. The earliest flying insects included grasshoppers and cockroaches, but these are not accomplished fliers. Even today, most species can only fly a few metres at most before they crash to the ground. The first true masters of the air were the dragonflies, swift and agile in flight. This must have given them huge advantages over earth-bound animals. They could swoop down on their prey from above and easily escape from predators such as amphibians. They could swiftly travel long distances to find food, or flee from approaching winter by migrating southwards. Colonising new habitats as they appeared would have been easy, and so it was flying insects that would always have been among the first to arrive as new islands arose from the oceans, and the first to colonise new ponds and lakes as they formed.

The final major innovation of the insects was the evolution of metamorphosis. Primitive insects do not change much as they grow, other than in getting bigger. The eggs hatch into miniature copies of the adults called nymphs, and these grow gradually larger by moulting. This is much the same as all the other arthropods, such as shrimps and spiders. Even today, many insects develop in this way: grasshoppers and crickets, earwigs, aphids and cockroaches, to name a few. Just a small number of insect groups undergo metamorphosis. In these, the eggs hatch into a grub or caterpillar, properly known as a larva, which looks nothing whatsoever like its parent. It is an eating machine, a mouth and digestive system contained within a flabby sac, designed for growth. It is usually not very mobile, has poor eyesight and generally weak senses. Most larvae rely on having been placed on or in a supply of food by their mother, and their job is to convert that food into insect tissue as quickly as possible. Once this job is done, the larva pupates, shedding its skin and turning into an immobile, helpless chrysalis or pupa. Inside the pupa, the tissues dissolve and are rebuilt from scratch. Wings form, and legs, eyes, antennae, a brain – all the body parts of the adult – are assembled. Once this is done, the adult bursts forth from the pupa, pumps up its wings to full size and is ready for action.

The job of the adult is to reproduce. Adult insects are sex machines: the males are 'designed' by evolution to find and seduce as many partners as possible, the females are designed to mate and then find a good source of larval food on which to lay their eggs. To do these things adult insects need to be highly mobile, and need to have well-developed senses with which to locate an appropriate mate and a place to lay eggs. Many adult insects, such as butterflies and mayflies, may live for just a few days; some for only a few hours. Some do not feed at all, or even have mouths to speak of. The advantage of metamorphosis is that it enables a division of labour between the different stages of the life cycle. The larva has a body that has been honed by evolution to allow rapid growth, while the adult has a completely different body design that is equipped for mating. More primitive insects, and other arthropods such as the arachnids and crustaceans, have to make do with the same body plan throughout their life, a kind of morphological compromise.

The advantage that metamorphosis gives to those insects that undergo it is illustrated by the enormous success of these groups, in terms of numbers of species. The four most speciose groups of animals on Earth are the beetles, flies, moths and wasps. The famous evolutionary biologist J.B.S. Haldane was once asked what his studies of evolution had taught him about the nature of God. He was an atheist, and replied, 'He must have an inordinate fondness for beetles.'

Today, of the roughly 1.5 million known species of plant and animal on Earth, 1.2 million are arthropods, of whom about one million are insects and, of these, about 800,000 are beetles, flies, moths or wasps. Of course there are millions more insects awaiting naming, should we ever get round to it. Take a swipe with a butterfly net in any tropical forest and you will almost certainly catch a handful of species that are new to science. The tricky bit is working out which ones have already been named and which ones haven't; for most groups of insects there may be only one or two specialists in the world who can work out which are which.*

For all of the last 500 million years almost every habitat on Earth has been dominated by arthropods, in terms of numbers of species and numbers of individuals. On land and in fresh water, most of these are insects. The dinosaurs came and went, followed by the great age of mammals, but through all of this there have been insects, swarming in mind-boggling diversity in the lakes, rivers, forests, grasslands and deserts, from the seashore to the top of the highest mountains.

---

* Sadly, funding for taxonomic work such as describing new species has shrivelled in recent decades, so such specialists are now hard to find. Soon there may be no experts left in many fields, so there will be no one to go to for help if you suspect you have discovered a species new to science.

Insects fill almost every conceivable ecological niche: they can be predators, parasites, herbivores or detritivores.* There is almost nothing of organic origin, alive or dead, that is not avidly consumed by insects of one sort or another. Some, such as clothes moths, can survive through their entire life cycle on foodstuffs that contain no moisture whatsoever. Others variously specialise in eating blood, wood, seeds, the tongues of frogs, dung, owl pellets, bacteria, leaves, algae, lichens, spiders, fungi and, of course, other insects. They vary in size from Bornean stick insects, which can grow to over thirty centimetres, to speck-like parasitoid wasps that weigh in at just twenty-five millionths of a gram. Their life cycles are staggeringly varied and unfamiliar. Leaf-mining flies may live almost their entire life burrowing inside a single leaf, while monarch butterflies regularly migrate 8,000 kilometres every year, from Canada to their hibernation grounds in Mexico and back again. Some, such as ants and termites, live in vast social colonies, with workers specialising as soldiers, gardeners or nurse-maids, while others, such as the death-watch beetle, may spend ten years alone and in darkness, slowly munching through the timber of a dead tree. Nymphs of the periodic cicada in North America spend seventeen years living underground, sucking on tree roots, before all emerging together to mate and die, while a fruit fly dashes through its entire life cycle in a fortnight.

Just as their life cycles are infinitely diverse, so their mating habitats are extraordinarily varied and often bizarre. While butterfly males use their beautiful wings to woo a mate, male scorpion flies offer piles of dried saliva as an enticement to females. Crickets, grasshoppers and cicadas sing to impress, while

* Detritivores are creatures that consume dead organic matter, including corpses, dead leaves and other plant material, and faeces. It isn't glamorous work, but somebody has to do it.

other insects such as moths release pheromones that waft for kilometres on the wind. Male stag beetles use their jaws to fight over mates, while male stalk-eyed flies (so named because their eyes are suspended on long, slender stalks on either side of their head) face off against one another – the one whose eyes are furthest apart winning the competition, and the female. Picture-wing flies perform elaborate dances for their mates, while fireflies and glow-worms use light-emitting bacteria in their bottoms to attract a partner. In dance flies, a group of small black flies that swarm in clouds above streams and ponds on a summer's eve, the usual pattern of sexual selection is sometimes reversed, with males providing expensive nuptial gifts to their mates – small dead insects carefully wrapped in silk – so that they are fussy about whom they choose, preferring large females who will have many eggs. To earn their gift, females have evolved swollen legs that make them look larger and thus fool the males. Some insects, such as earwigs and shield bugs, care for their offspring, even sacrificing their life to save that of their young. Corpse-eating carrion beetles look after their offspring, but if they have too many they casually consume the surplus, ensuring an adequate supply of dead meat for the remainder. No doubt there are many more marvellous behaviours that remain to be discovered, if we ever find the time to look.

In terms of numbers of individuals, insects rule supreme. A single leaf-cutter ant nest in the forests of South America can house four million ants. At any one point in time there are currently thought to be very roughly ten million trillion individual insects alive on Earth (Ethel got a bit carried away in 'The Insects' World'). Whatever way you look at it, we are seriously outnumbered. Some pest insects, such as aphids and house flies, are perhaps more common than ever because of the food we unwittingly supply for them. But most insects are declining, and many thousands of species have already gone extinct. As the most famous entomologist alive today, E.O. Wilson, once said, 'If all mankind were to disappear, the world would regenerate back to the rich state of equilibrium that existed ten thousand years ago. If insects were to vanish, the environment would collapse into chaos.' Insects *are* vanishing. By every indicator we have, the bulk of insects are declining. Butterflies, bees, dragonflies, grasshoppers and the countless species that live only in the dwindling rainforests are disappearing one by one.

In the minds of many, conservation is all about giant pandas, ospreys, tigers, rhinos and blue whales: large, charismatic, furry or feathery creatures, often living on the other side of the world, glimpsed only in television documentaries. What few people appreciate is that the vast majority of life on Earth, in terms of both numbers of species and numbers of individuals, is made up of insects and other arthropods, and that many of them are just as important, fascinating and worthy of our interest and of conservation as the larger creatures. Indeed, whilst the extinction of the giant panda would be terribly sad, it would not have any knock-on consequences. There would perhaps be a tiny bit more bamboo in a forest in China. In contrast, the little creatures that live all around us are absolutely vital to our survival and well-being, yet we generally pay them little heed unless they annoy us.

The various flowers in my meadow need bees, hoverflies, butterflies and beetles to pollinate them, and many of those same insects fly out of the meadow to pollinate the sunflowers in the neighbouring fields and the peaches, apples and tomatoes in my small garden by the house. The wild flowers and my vegetables also need a healthy soil in which to grow, and so depend upon the springtails, woodlice, worms and millipedes that live in it, recycling nutrients and aerating the soil. Without predators such as ladybirds, lacewings and rove beetles, and parasitoids such as many small types of wasps and flies, herbivorous insects would run amok, wiping out their preferred host plants and destroying the ecological balance of the meadow. Without grasshoppers, flies, crickets and moths, the birds and bats would have no food. Ugly or beautiful, it is the little creatures that make the world go round. We should celebrate and appreciate them in all their wonderful diversity.

CHAPTER THREE

Chez les Newts

3 September 2007. Run: 41 mins 31 secs. My legs felt heavy today, I must reduce my nightly cheese intake! People: none. Dogs: 3. Butterfly species: 9 – a poor haul; the insect year is drawing to a close. An unseasonably late glow-worm was casting her magical green light on the patio here last night, but I can't see her here now. It is hard to believe that these little snail-eating beetles have evolved to culture phosphorescent bacteria in their bottoms as a means to attract a mate – there must be an easier way! Great year for fruit; I snatched huge, glistening blackberries from the hedgerows as I ran – my fingers are purple from the juice – and my decrepit, lichen-encrusted peach trees in front of the house are for once laden with aromatic, golden fruit.

Amphibians and I have something of a history. If you happen to have read *A Sting in the Tale* you may recall the unfortunate fate of my frostbitten Chinese painted quails, and the accidental electrocution of my tropical fish. My childhood misadventures with my menagerie of pets also resulted in the demise of various amphibians, some of which haunt me to this day. And so it is perhaps just as well (at least for the amphibians) that I chose to focus my adult career on insects.

From a very young age I kept newts and common toads in tanks in my bedroom, and this went atypically well. The toads in particular made great pets, seemingly taking to captivity and providing great entertainment by hoovering up mealworms with their extending, sticky tongues. When I grew bored of them, or ran out of mealworms from the supply that I bred in a box under my bed, I could simply release the toads back into the garden. However, I longed to have some more exotic amphibians, and eventually I badgered my parents into buying me a pair of North American leopard frogs for Christmas: attractive, bright-green frogs with (as you might guess from the name) a profusion of black spots. I filled one of my glass fish tanks with piles of stones, peat, some plants and a small pond, to make an attractive home for them. It looked great and the frogs settled in well, but after just a few weeks their energetic hopping about caused one of the piles of stones to topple; I came home from school one day to find them both squashed.

Undeterred, a year or so later I saved up my pocket money and bought an axolotl. It was a bizarre and wonderful creature. Axolotls are effectively giant tadpoles, reaching sexual maturity while still having the external fluffy gills and purely aquatic habit normally associated with the immature stage of newts and salamanders. They are found in the wild near Mexico City, although they are now critically endangered because of pollution and urban development. Fortunately there are plenty in captivity, particularly in labs, where they are kept to study their unusual ability to regenerate lost limbs. I had a large tank half-filled with water in which three baby red-eared terrapins lived, and I thought this would be perfect for the axolotl. It was both larger and faster than the endearing little terrapins, so it did not occur to me that they would do it any harm. I released the axolotl into the tank, watched them all swimming around for a little while – all seemed peaceable enough – and went to have my tea. I came back a while later to inspect my new pet, only to find that it had been consumed almost entirely by the terrapins, which turned out to be far more ferocious than their appearance suggested. The three of them had hauled themselves out on to a rock under their heat lamp to digest their huge meal, and only the head and some of the spine of the axolotl remained – it was clearly not going to be regenerating. I was distraught, and feel awful to this day when I remember the incident.

In my early teens I acquired two young Argentinian horned toads: fabulously devilish-looking beasts, multicoloured in green, orange and black splodges, with capacious mouths. When fully grown, they can reach nearly thirty centimetres in length and are said to be able to swallow a rat. They thrived for a while, until one swallowed the other. I found the pair of them dead, the aggressor having presumably choked on its meal, with the feet of the victim still protruding from its mouth.

Undeterred, I tried keeping a White's tree frog. This was a charming little creature, turquoise-blue with huge, sticky fingertips. Unfortunately I did not appreciate the need to add a calcium supplement to its diet of insects, and it developed rickets; its leg bones became flexible, so that it had difficulty hopping. I quickly purchased some calcium and started sprinkling it on the frog's food, but its limbs hardened in deformed shapes. Nonetheless the frog survived and managed to get about, albeit somewhat awkwardly. It lived for a while, until one day I must have left the lid to its tank slightly ajar and it somehow squeezed out. I searched high and low for it, but to no avail. It was nearly two years later that I found the hapless creature in the tip of an old pair of trainers, where it had clearly decided to hide. It had presumably died of dehydration (or perhaps the smell), and its body had mummified.

That was my last attempt to keep amphibians as pets – clearly they were too tricky for someone as incompetent as me to look after. Sadly, as we shall see, this was not the last time I was to accidentally inflict misfortune on these fascinating creatures.

In the late summer of 2003, a few months after taking over

ownership of Chez Nauche, I took a party of twelve students from the University of Southampton down there to blitz the renovations. There was a daunting amount to do on the old place to make it even remotely habitable, and having a small army of volunteers seemed like a good idea. In exchange for two weeks' labour I offered to pay for their travel costs and provide them with as much food and cheap red wine as they could consume (rather a lot, as it turned out). It was an ill-thought-out plan, for only one of them had any building experience; and I had not anticipated the difficulties in coordinating and overseeing the activities of a gang of novices armed with power tools when I myself had little idea what I was doing. It is a small wonder there was anything at all left of the old place by the time we had finished, and even more surprising there were no major injuries.

On the first day I divided the students up into work parties. One gang demolished a section of old leaky roof, hurling down the broken clay tiles while perched like monkeys on the old, rotting roof timbers. A second group knocked out the rotten windows and door frames, so that there was soon broken glass all over the floor, while a third party set to demolishing some internal walls. I ran around like a headless chicken trying to give advice and avoid catastrophes. The old building echoed to the sound of crashing tiles, the thump of sledgehammers and the ear-splitting whine of the angle-grinder. To my considerable dismay, in all the chaos a lovely old ceramic sink and some ancient wooden-framed classroom slates that I had found in the loft were smashed. If there is a French equivalent of English Heritage, they would have been horrified to see what was going on. On the other hand, progress was pretty rapid, and great piles of broken tiles, old bricks and rotten window frames began to accumulate along the front of the house. It was hot work, and the rust-red dust stuck to our sweaty skin, so that we were soon utterly filthy.

There was a rough, dry hollow near the edge of the drive, some thirty metres uphill from the house, and it seemed like the perfect place to dispose of all the waste. We shovelled the rubbish into a wheelbarrow, pushed each load laboriously up the hill and tipped it into the hollow with a clatter of dust, and slowly the hollow disappeared. This was something that I was eventually to regret, for it had not occurred to me that this hollow might not always be dry.

At the end of each day we took it in turns to shower around the back of the barn, dousing ourselves in icy-cold water from a hose that I had rigged up. It was glorious standing naked in the long grass, washing off the fine red dust in the evening sunshine. We cooked on a camping stove under the stars, and then sat late into the night drinking beer and wine, nibbling extraordinary quantities of cheese and listening to the eccentric musical tastes of James Peat, whom some of you may remember from *A Sting in the Tale* as the PhD student who made his own trousers. He had brought along his iPod, something that in 2003 seemed improbably, almost magically, advanced. Since the house was far from habitable, we slept in a cluster of tents in the field, and at times it felt as if we were forming some sort of hippie commune. Indeed, if ever I was away buying supplies, I am told that James would often take the opportunity to wander around the farm naked, but he seemed uncharacteristically self-conscious when I was around.

The wall-demolition group made quick progress, as most of the internal walls were flimsy partitions built from fibreboard by Monsieur Poupard. They served little purpose other than to divide up the internal space into a series of dark, dank holes. The only substantial internal partition was an ancient construction of sturdy vertical beams joined by oak laths plastered with straw, dung and mud, which had divided Monsieur Poupard's bedroom from a small, earthen-floored barn beyond. The wall was built on to the bare earth, and the bottoms of the wooden posts had rotted beyond repair, so it had to be removed before it collapsed. James was one of the gang on wall demolition, and he attacked this wall with gusto and a large sledgehammer. The whole house was thoroughly damp, thanks to the leaky roof, but nonetheless James was surprised to spot movement amongst the piles of crumbled mud and broken, woodworm-ridden beams that he was creating. He called me over, and we disinterred a very sizeable newt. A little searching revealed two more wriggling amongst the wreckage, a little battered, but still more or less in one piece. We washed them off in a bucket, exposing them to be stunningly beautiful marbled newts, their velvet-black skin punctuated with bright-green irregular spots. They are relatives of the great-crested newt found in Britain, but how they came to be living in my house was unclear. I guess it says a lot about just how damp the place was.

After completing the demolition works, we began repairing the damage. I attempted to hone my primitive plumbing skills to provide a working toilet. Alistair, a mature student with previous building experience, was the only one of us who had any idea what he was doing. He could slap on the rough rendering needed to hold together the old stone walls at an impressive rate of knots when the mood took him, but he started drinking at lunchtime and was far more interested in chatting up one of the female students than he was in plasterwork – for which I could hardly blame him, given that he wasn't being paid – but this meant that his rate of progress was erratic at best. The roofers started trying to rebuild the old roof they had removed. A guy called Dave tried to fit the new windows that I had had delivered, a frustrating task since none of the holes was quite square, and none of the metric windows of quite the right size for the holes. James mixed up concrete and laid a new floor. Ben attempted some wiring.

A slightly scatty girl named Callie spent most of her time trying to catch the numerous wall lizards by luring them into beer bottles – an eccentric and wholly unsuccessful endeavour, intended to stop them becoming accidentally entombed by Alistair's rendering. It did nothing to help progress.

Midway through our stay, the French hunting season began. We were woken at dawn by the sound of gunfire. It sounded as if war had broken out, for there was a near-constant crackle of gunfire. The French are famous for their enthusiasm for shooting and eating almost anything that moves, and this is no myth. The French countryside is full of wildlife, but any animal larger than a blackbird tends to be nervous and secretive, for it is in constant peril of being shot. French rabbits are scarcely ever seen, although their pellets and digging activity suggest that they are common at Chez Nauche. Unlike their British cousins, which brazenly graze by the side of busy roads in broad daylight, French rabbits are strictly nocturnal, for good reason. If they were not, they would soon find themselves on a dining table.

We tried to ignore the rattle of shots and carried on with our work. On that morning I was helping the roofers, the three of us perched on the ridge of the building, from where we had a fine view in all directions. The wack-wack bird, which had been calling every day previously, had sensibly fallen silent. We spied a party of hunters heading along the green lane at the northern boundary of the meadow. There were five of them, each armed with shotguns and with several dogs gambolling excitedly around their feet, ready to collect the carcasses of any animals that their owners shot. As the hunters reached the nearest point along the lane to us, perhaps 150 metres to our north, one of the dogs put up a red-legged partridge. These are lovely birds with cream-and-chestnut barring on the sides of their breast, and this one was particularly plump. It took off with a clatter of wings and a squawk, heading towards us. One of the hunters swung his gun to follow it and fired two shots in our direction. After a second's pause, lead shot rained down upon us, its force largely spent, but nonetheless causing us to scuttle for cover behind the chimney pot. The bird must have been clipped by one of the shots, for it crashed to the ground in a puff of feathers, but then picked itself up and ran for it. One of the hunters pushed through the hedge and set off in pursuit.

I was hopping mad, both at being shot at and because the hunters were clearly intent on exterminating wildlife on my intended nature reserve. I shinned down the ladder and ran towards the hunter, shouting and waving my arms. I'm not sure what I was shouting, but in the heat of the moment I had lost all my flimsy command of French, so I'm sure it made little sense. As I got closer I realised that the hunter was a stout woman of late middle age, wearing one of those hunting, shooting and fishing waistcoats that has hundreds of pockets for ammunition, knives, and so on. She even had a bandolier of shotgun cartridges slung diagonally across her chest. What I didn't immediately notice was that her tweed cap was covering a pair of ear-defenders, which would explain why she didn't hear me shouting. She was still intent on her prey. Her dog had run ahead, and suddenly the partridge took to the air again, rather wobbly this time, and she swung her gun and shot it. It was at close range, and the creature was obliterated. I was shouting very loudly indeed by this point, and finally she heard me and swung her gun in my direction. For a moment I thought I was going to get the second barrel. Thankfully she was able to tell the difference between an angry Englishman and potential dinner at such close range and didn't pull the trigger. We had an interesting and somewhat heated debate, during which I attempted to marshal enough French to explain my indignation, while she shrugged repeatedly and seemed bemused by my anger. The dog, meanwhile, had retrieved what remained of the partridge and in a gesture of goodwill she offered me the ragged, dripping, shot-laden corpse. I declined, and we may or may not have agreed to differ as to whether it was acceptable to pepper someone with lead when sitting on the roof of their own home. She departed to resume the slaughter, and no doubt later to recount the tale of the mad Englishman who had moved into Chez Nauche. The next day I nailed up '*Chasse Interdit*' signs along the farm boundaries.

There were many other moments of excitement on that trip. At intervals my power tools burst into flames, which was none too helpful. We were still using some of the rudimentary ancient wiring, including cloth-covered (rather than plastic) wires – something I'd never seen before – and we eventually realised that the problems only occurred with two particular sockets. I later discovered that, due to a quirk in the wiring, these sockets were delivering twice the normal voltage: highly dangerous to both power tools and humans.

By the end of a fortnight I had a house with a waterproof roof, a few windows (and rather a lot of gaping holes where windows were needed), a working toilet and electric lights. It all looked dreadful, but it was slowly approaching being habitable, in the loosest sense of the word. We had got through an awful lot of cheese and red wine, but it seemed worth it.

It wasn't until the following spring when, accompanied by my dad, I made my next DIY trip to Chez Nauche that I realised the mistake I had made in filling in the hollow. I noticed that the rubble we had thrown in was now inundated with water, which fed in from a ditch running down the side of the drive. What had been a dry hollow in summer was clearly a pond in winter and spring. Newts have to have a pond to breed in – they go to them to mate and spawn in early spring, and then leave them again in the summer. The truth dawned uncomfortably upon me. The hollow that we had more or less completely filled with rubble was in fact a temporary pond – in winter it filled with water, but by summer it was dry. There were no other ponds in the thirteen hectares at Chez Nauche, so I had successfully destroyed the only breeding habitat for these lovely creatures. Given that my aim in taking over the farm was to create a wildlife sanctuary, this wasn't an auspicious start.

The obvious thing to do would have been to dredge out the rubble. I cannot recall why I initially decided against this. Instead, I hatched a plan for a much larger replacement pond, which I would create by damming the stream flowing from the small spring in the meadow. I reasoned that I could dig out a substantial lake in the valley below the spring. I envisioned an expanse of wind-ruffled water sparkling in the sunshine and reflecting the clear-blue Charente sky, with fish turning, perhaps the odd heron, and many happy newts frolicking in the shallows.

My initial thought was that I would perhaps get someone with a digger to scoop out the lake, so I asked around among my limited local contacts. I eventually found a chap by the name of Marcel who came to look at the spring and, if I understood him correctly, declared that he could create a fine lake. However, his quote for the work, when it arrived by post a few days later, was for 7,000 euros, which would have made it the world's most expensive newt home. Plan B was to dig it out by hand. After all, the Suez Canal was dug by hand – how hard could it be to make a small lake? I figured that the spoil heap could be used to create a mighty dam across the valley. And why stop at one lake? I could have a whole series, with the water spilling attractively over a series of weirs from one to the next. In my head, it was a marvellous scheme.

So it was that, on subsequent trips to France, whenever I had a spare moment I would dig. There are always lots of other things to do, making the house more comfortable and looking after the meadow, so the digging wasn't a top priority and it went slowly. I initially made three dams about fifteen metres apart, each strengthened by several wheelbarrows full of stone rubble carried down from the partially demolished barn. The soil in the valley is heavy clay and full of flints, so it is exceedingly difficult to penetrate with a spade, and when it dries in summer it becomes as hard as concrete. The only way to make any inroad at all was to hack at it with a pickaxe, and then shovel up the loosened soil – blistering and back-breaking work.

By 2008, despairing of ever seeing the mighty expanses of water that lurked only in my imagination, I decided to organise another work party. At this stage I was at Stirling University, and I invited down a group of friends to dig. It might not seem like the most attractive prospect, travelling 1,600 kilometres to dig a hole in the ground, but one should never underestimate the appeal of the offer of unlimited supplies of French cheese and wine. Five of them volunteered, all staff from the university, and so we formed a chain gang, hacking away at the hard earth and taking it in turns to bring more stone down from the barn. We focused our efforts on one lake; the dam slowly rose, and the hole widened. It was September, and the sun shone from a cloudless blue sky, as it tends to do at that time of year. It became hellishly hot in the hole, sheltered as it was from any hint of a breeze. After a week of dusty toil and blistered hands we had a cavity perhaps two metres deep, six metres wide, and nine metres long. It still wasn't really going to be enough to go boating on, but it was considerably larger than the hollow I had foolishly filled in up near the house. It should certainly be big enough for quite a lot of newts. However, it had been a dry summer and not a drop of rain had fallen during our stay. The trickle of water from the spring petered out before the hole, sinking into the parched soil, so I guessed it would not fill up until the autumn rains set in, as they usually do. We shut up the farm for the winter and returned home.

The following spring, on my next visit, I ran down to inspect the pond. I was sure it would be full, with perhaps a few frogs and newts, water beetles, pond skaters, and so on. As I jogged down the hill I recalled how quickly the pond I had dug as a child in our garden in rural Shropshire had been magically colonised by a host of fascinating aquatic life. Imagine my disappointment to find the hole largely unchanged. There was a tiny puddle in the bottom, perhaps forty-five centimetres across, and one solitary frog, which ineffectually attempted to hide in the few centimetres of water. It had clearly not been a very wet winter, for the meadow grass was shorter than usual, but this was not the cause of the problem. A mole had, quite literally, undermined all our hard work. The lower face of the dam was dotted with molehills, and the steady trickle of water coming down from the spring was pouring straight through the mole tunnels and under the dam. Why the mole should have chosen to dig here, in the bottom of a stony hole in the ground, when there was a whole meadow of softer, worm-rich soil for it to burrow around in was beyond me.

I battled with that mole for the next two years. I puddled clay to a thick, smooth consistency and poured it into his tunnels, hoping that it would dry and set hard. The mole was undeterred and every morning new molehills appeared. I rammed rocks down all of the holes, stamping them in with my feet and sealing around them with more clay, but the mole just popped up elsewhere; every morning two or three more molehills would announce his stubborn resistance to my attempts to drive him away. In desperation I mixed concrete and lined the inside face of the dam with it, but before it could set each evening the mole pushed holes through it, leaving a pile of crumbled cement and newly excavated soil. I began to empathise with the American soldiers in Vietnam, ineffectually trying to flush an unseen enemy from its underground network of tunnels and bunkers. For all my efforts, the hole remained just that, home to a solitary frog and, of course, a mole.

It was not until the spring of 2012 that I finally won this battle. Perhaps the mole simply died of old age and, if so, I imagine that he had a smug, satisfied smile upon his furry pointed face when he went. The weather may have been what turned the tide, for the late winter was exceptionally wet, and the water from the spring flowed as strongly as I have ever seen it, perhaps flooding the mole out of his network of tunnels beneath the dam. The torrent of water had washed mud down, silting up the deepest part of the hole and perhaps blocking up the tunnels, although there were still a few leaks. Whatever the cause, when I visited in May 2012 the pond was full of water, nearly reaching the top of the dam. What was even more exciting, it was alive with creatures. Several frogs leapt from the banks as I approached, plunging into the cool depths. Whirligig beetles gyrated on the water surface, perhaps having flown up from the pools on the meandering Transon nearby. Pond skaters skittered about, disturbed by the frogs, and water boatmen rowed their way jerkily along beneath the surface. Damp-loving ragged robin had appeared from nowhere and was flowering on the bank, offering a perch for dozens of damselflies, including both common red and blue-spots. There were no demoiselles, for they seem to prefer running water. But there was a blue chaser dragonfly perched on a dead thistle stem hanging over the water, a lovely species with a fat blue body with a powdery bloom like a plum, and powerful, fast flight. I couldn't resist it, and quickly stripped off and jumped into the pond, causing quite a stir amongst the whirligigs. There wasn't really room for much of a swim, more of a muddy wallow, but it was wonderful until I noticed a lot of coypu faeces floating around me, which slightly took the edge off the experience.

In *The Field of Dreams*, a strange and rather dull movie about baseball, Kevin Costner is advised that 'If you build it, they will come.' I'm not sure this is generally useful advice, particularly if moles are involved (so far as I recall, no moles attempted to thwart Kevin's efforts in the film, although it might have been more interesting if they had), but it is certainly true of ponds. I would recommend that anyone who has a spare corner of their garden should consider installing one, because there is no doubt that it will soon be teeming with insects and, if you are lucky, with a few amphibians too. It is the one single addition to any garden that will make the biggest difference in encouraging wildlife. Sadly, the only creatures that have not yet arrived at my pond are the newts that motivated its construction, but I live in hope. Every spring I check for them with my boys, using our pond-dipping nets to dredge out any that might be lurking in the depths, but as yet without luck. I am sure that, given time, they will come.

CHAPTER FOUR

Mating Wheels and Sexual Cannibalism

9 May 2008. Run: 39 mins 6 secs. A little faster today – perhaps the marathon training is paying off! People: 2 – a rare sighting of young people in Épenède, a little girl walking to the postbox with her mother. Dogs: 4. Butterfly species: 11, including dainty wood whites fluttering demurely in the shaded lane along the south-west edge of the meadow. I also came across a hedgehog out for an early-morning snuffle, or perhaps on its way home after a busy night of chomping worms. I love their rolling gait, reminiscent of an old man with a gammy hip on the way back from the pub.

Dragonflies are amazing but primitive insects, and I was thrilled to see them occupy the new pond at Chez Nauche. They have not changed much since the Carboniferous, 320 million years ago, when they were the largest animals in the air. Back then, some species were well over sixty centimetres across, the top aerial predators of their day, and they remained kings of the sky for 100 million years until displaced by the pterosaurs. Dragonflies snatch their prey in mid-air, scooping them up with their forward-curved, spiny legs, which form a basket beneath them as they fly. When hunting they can accelerate to well in excess of fifty kilometres per hour, much faster than most of their hapless prey, which includes flies, bees, wasps, butterflies, damselflies and even other dragonflies.

Dragonflies have the largest eyes in the insect world, with up to 30,000 facets, and wrapping entirely around their head so that they can see in all directions at once. Close up, their eyes have a multi-coloured iridescence, a little reminiscent of the colours on a compact disc, but with a three-dimensional, almost holographic, effect.

There are many fascinating aspects to the biology of dragonflies. They lay their eggs in fresh water, and their nymphs are themselves voracious predators. They are squat and brown, ugly ducklings that show no signs of the beauty that is to come in adulthood. They feed with a mechanism similar to the gruesome alien in the film of the same name; they are ambush predators, waiting motionless until a small fish or tadpole swims near. Then, in the blink of an eye, their telescopic jaws unfold forward from their face and their prey is punctured by their sharp mandibles and drawn back to be consumed. When fully grown, the nymph climbs out of the water, generally using a reed stem, and the adult bursts out from within, leaving behind a papery brown husk.

The adults are quite long-lived by insect standards, some spending several months on the wing, searching for a mate and laying eggs. Their courtship and mating are peculiar even by the standards of the wonderful world of insect sex. The male dragonfly has a pair of claspers at the tip of his long, thin body, which perfectly fit the base of the neck of the female of his species, but will fit no other. Male dragonflies spend much of their time searching for a female, or guarding a resource – a pond – that they know females need to visit to lay their eggs. If they see a female they will attempt to grab her by the neck, either in mid-air or by waiting until she has perched to lay an egg. They will happily grab newly emerging females who are not yet able to fly, or will attack mating couples, attempting to tear them apart and grab the female for themselves. The females usually try to avoid male attentions, and will take evasive action if they have the chance.

Once a male has captured a female, he will not readily let her go. In many species the male's claspers are armed with spines, which actually stab into the head of the female, making it extremely hard for her to escape or for another male to separate them. Couples may stay locked together for days, flying around in tandem, with the male in front, the female being towed behind by the scruff of the neck. It may sound undignified, but they look rather beautiful together.

Before they can actually mate, there are various logistical problems that they must overcome, and that require the cooperation of both parties. First, the male's testes are positioned near the tip of his body, which is now behind the female's head. Unusually, his penis is tucked up near the front of his abdomen, just behind the wings. If he has any sense, he will have carefully transferred a packet of sperm from the testes to the base of the penis by bending his abdomen double before he grasped hold of the female. If he did not, he is wasting his time. For them to mate, the female has to curve her own abdomen forward and underneath her head, so that her genitalia at the tip come into contact with the male's penis. At this point the couple's bodies form a rough circle or, more precisely, a rather romantic, lopsided heart shape, sometimes known as a mating wheel. Once her reproductive parts are in place, the male uses his penis to scrape out the receptacle in which the female stores sperm, in an attempt to remove all the sperm from any previous suitor. To this end, male dragonflies have penises that, in different species, variously resemble ice-cream scoops or scrubbing brushes. Once this sticky job is complete, the male transfers his own parcel of sperm. Having done so, he hangs on to the female for as long as he can. If he lets go, she is sure to be grabbed by another male and his sperm will then be unceremoniously removed, so if he wishes to have offspring he must hang on until she uses his sperm to fertilise her eggs. So it happens that, for much of the summer, most adult dragonflies are in pairs. They will hunt, sleep and she will lay eggs while locked together. Once she has laid eggs, he can be fairly sure that he has fathered at least some of the offspring.

You may wonder why the female would be willing to mate after being grabbed by a male without courtship, and after trying to evade him and reject his advances. The answer is presumably that she has to mate with someone. Having done her best to escape, she has established that this particular male must be fairly fast and strong, and so hopefully her sons will inherit his vigour.

There is something of a puzzle as to why dragonflies have evolved such a bizarre and complex way of mating. The explanation may be that it came about as a means of avoiding being eaten by their intended mate. In most mammals, males tend to be bigger and stronger than females; they need to be able to fight for access to females. Hence male red deer, the stags, are much more powerful and have much larger antlers than the females. In contrast, in insects it is generally the females that are larger; the bigger the female, the more eggs she can produce, whereas in males high mobility (to find mates) is usually more important than size. Hence natural selection has favoured voluptuous females and skinny, nippy males.* As a result, sex in many predatory insects and also spiders is fraught with danger for the male of the species, for the females are often the larger and more powerful sex. Male dragonflies have largely overcome this problem by swooping on the object of their desire from above and behind;

---

* There are of course exceptions to this general pattern; insects are such a vast group that there are always exceptions. In stag beetles, for example, the males fight each other for females, and so they tend to be the larger sex, and also have greatly enlarged jaws for fighting.

once he has firmly grasped her by the neck, he is safe from her powerful legs and jaws.

Male mantises have not yet come up with an entirely adequate solution to this problem. It may be that mantises evolved relatively more recently, about 100 million years ago, and so have had less time to come up with an answer than the dragonflies. The female mantis is considerably bigger and stronger than the male, so she is readily able to overpower and eat him, should she prefer a snack to sex. For the amorous male, gauging the mood of his potential mate as accurately as possible could not be more important; it is a matter of life or death.

It has long been known that female mantises commonly eat their mate during the act of copulation – the famous French entomologist Jean-Henri Fabre described this in the 1800s. Some have argued that it is an essential part of mating. Indeed, once the male's head has been removed – this is usually the first body part that the female consumes – his mating seems to become more enthusiastic, his body pumping vigorously. Of course his body might as well give it everything at this point, as it isn't going anywhere afterwards. Indeed, the male often doesn't get as far as copulating before he loses his head. Bizarrely, the male can still successfully initiate copulation *after* his head had been removed. In a series of odd experiments carried out by K.D. Roeder of Tufts University, Massachusetts, in the 1930s he demonstrated that successful mating can in fact take place when neither partner has a head, though perhaps this is of little relevance to what happens in nature.

It has been suggested that it may be in the male's interests to be consumed during copulation, for he is providing a large meal to his partner that may enable her to produce more of his offspring. However, this is only likely to be so if his chances of finding another female and mating again are slim; otherwise he is giving up all chance of mating again for the benefit of providing one meal to his current partner. Males of most animals are not so self-sacrificing.*

Some years ago I had the chance to study this for myself. I had gone on a cheap package holiday with my wife to Gambia, a country with an immense diversity of wildlife. The area around our hotel was almost dripping with mantises of various species, and their large egg masses, known as oothecae, were stuck all over various ornamental shrubs and could even be found on the walls of the hotel. I'd always been intrigued by the mating habits of mantises, and this seemed like an ideal opportunity to get hold of enough to do some experiments. I brought some oothecae back with me, and in the dingy basement of the biology building at the University of Southampton I set about rearing them.

Many of the oothecae produced only tiny parasitic wasps, the females armed with enormous ovipositors far longer than their bodies. These they clearly use to stab into fresh oothecae, laying their eggs deep inside, where they hatch and consume the eggs of the mantis. I pickled them in alcohol, and to this day have not got round to checking whether they belong to a species known to science. Fortunately some oothecae had escaped the depredations

---

* Tropical orb-web spiders are a fascinating exception. Male orb-web spiders are tiny compared to the females, and it has long been known that they are frequently consumed by their mates before, during or after copulation. It has recently emerged that the male genitalia (strictly their palps, which are used to transfer sperm into the female) often snap off inside the female during sex. This has the advantage for the male that it permanently blocks her reproductive tract, preventing her from mating again and so ensuring his paternity of her offspring. Of course the disadvantage is that he is now a eunuch, and he may as well get eaten since he has no way of ever mating again.

of the wasps, and dozens of tiny mantis nymphs emerged from these. Rearing mantises is problematic, as they are enthusiastic cannibals from day one. If they are housed together, a group of a dozen small mantises quickly becomes one bigger mantis, so they each have to be housed and fed individually, on fruit flies to start with, moving on to house flies and then small crickets. Luckily one of our technical staff, a shambling bear of a man called Keith, absolutely loved mantises and gave me heaps of help. Eventually, after an awful lot of work, we had several hundred splendid adult mantises, each about ten centimetres long. We kept them for another week or so to make sure they were mature, and then I began my experiments.

I wanted to watch exactly what happened during courtship, and to see whether cannibalism only occurred when the female was hungry. I also wanted to see whether cannibalism was reduced when mating took place in a spacious environment with lots of twigs and cover for the male, rather than in a bare cage. I spent many happy hours watching my mantises, aided by a succession of undergraduate project students.

Typically, and understandably, the male mantises tended to approach the females extremely slowly. Mantises can move almost imperceptibly by swaying gently, as if they are a frond of vegetation being wafted by the breeze. The impression is helped by their camouflage; most species are green or brown, so that they blend in well with the vegetation in which they live. A few tropical species are spectacularly beautiful, being pink or white to match the exotic orchids on which they wait to catch prey. The French species found in my meadow is a vivid, glossy green, the colour of wet summer grass; my Gambian mantises were a twiggy brown. With each tiny sway the mantis moves slightly more forward than back, and so edges towards its target: often prey, sometimes a mate.

When courting, the males proved to have two distinct strategies. Some attempted a front-on approach, edging more and more slowly towards the female, the closer they got. When they were just out of reach of her powerful forelegs, they started to display. The exact display varies from species to species; mine engaged in something akin to belly-dancing, with the male gyrating his abdomen, curling it forwards and backwards, wiggling it from side to side, and sometimes managing both at once. The female watched, inscrutable. Or at least so it seemed to me, but presumably the males were watching her to see if she was impressed. What kind of mood was she in? Did she look amorous, or peckish? If he felt the moment was right, he would pause, crouch and in a single bound would leap on to her back, pivoting in mid-air as he did so, like a vaulter at the Olympics.

Other males approached the females from behind. In this case a male would typically approach slowly at first, but – in marked contrast to males approaching from the front – he would accelerate as he got closer, sprinting the last few centimetres and then hurling himself on to the female. So far as I could tell, these females had no idea he was there until rudely mounted from behind.

Once on the female's back via either strategy, a male is not within easy reach of the female's forelegs, but he is not safe. She invariably struggles when he first lands on her, presumably weighing up her options. If he is lucky, she may decide to accept his attentions. If not and she continues to struggle, it is only a matter of moments before her superior strength prevails and he becomes dinner. This might occur if the female decides that the male is of inferior quality – an extreme form of rejection – or it might be just because he caught her at a bad time. Sometimes she lets him mate, but then decides that she needs a snack halfway through and that is when, typically, she bites off his head.

The results of our studies were pretty clear. At no point could we discern any sign that the male was a willing victim of canni-balism. Their approaches to females were cautious, and after mating they would either leap off and run for it or edge discreetly off her back, keeping well away from her front end. When confined with a female in a simple, undecorated cage, the males were much more likely to be consumed, usually without mating, than when there were plentiful twigs and leaves. When confined with a hungry female, their chances were not great. But when the female was well fed and they were confined together in a twig-filled cage, rates of cannibalism were very low, with males surviving more than 90 per cent of matings. I have since repeated this with French mantises from my meadow, and these Gallic lovers show no more signs of wishing to be eaten by their partners than did their Gambian counterparts. None of this tells us for sure how common sexual cannibalism is in nature, but it suggests that it may be nowhere near as common as once supposed.

Sex in insects is fraught with conflict. A meadow in summer is a seething mass of sexual adverts, courting couples, brutal rejections, conquests and copulation, a mad rush to reproduce and ensure there are offspring to carry on the line when the season ends. The song of the birds and crickets, the flash of a butterfly's wing, the swarming of dance flies on a summer's eve – all are driven by the evolutionary imperative of reproduction. However, as we shall see in chapter twelve, sexual intrigue is not confined just to animals.

CHAPTER FIVE

Filthy Flies

17 July 2008. Run: 38 mins 17 secs. People: one friendly old lady on her bicycle. Dogs: 5. Butterfly species: 12, the highlight of which was a silver-washed fritillary restlessly nectaring on brambles, a powerful creature with black-spotted rusty wings, the hue of dry, late-summer bracken. A cloudless blue sky once more, and it is already warm at 8 a.m. As I am recovering from the run, gasping and damp, I can see a female green lizard burrowing into my heap of building sand, preparing a nest for her eggs. In the distance, the wack-wack bird is calling plaintively from the south-east.

> *God in His wisdom made the fly*
> *And then forgot to tell us why.*

Ogden Nash, 'The Fly'

> *The hand is quicker than the eye is,*
> *But somewhat slower than the fly is.*

Richard Armour, 'Inscription for a Fly Swatter'

Flies are a group of insects that has few fans, which is a great shame. God may not have told us why he made them, so I shall give it a stab. Flies are a huge and diverse group, comprising about 240,000 known species, roughly one-fifth of all known life on Earth. Following the logic of J.B.S. Haldane's quip that

God must have 'an inordinate fondness for beetles', I would add that He must also have a penchant for flies. They are found in more or less every terrestrial and fresh-water environment on Earth, from the deserts to the Arctic, from the sludge at the bottom of garden ponds to all but the top of the highest mountains. The larvae of one species can survive temperatures above boiling in the trunks of the desert cacti in which they live. Flies are a staple food for many thousands of animals – at Chez Nauche they are eagerly consumed by swallows and martins, pipistrelle bats, lizards and toads, mantises and many other fascinating creatures. Millions of birds migrate each year to the Arctic Circle to breed, so they can feed their young on the countless midges that emerge in the brief Arctic summer. The larvae of many flies are vital in helping to break down rotting organic matter, recycling the nutrients so that they become reavailable for plants to use in growth. Some flies are strikingly beautiful, clothed in metallic green, blue or gold; the large furry hoverflies that mimic bumblebees are almost cuddly. Picture-wing flies have intricately patterned and coloured wings, like miniature stained-glass windows. On the muddy edges of my pond, bright-green dolichopodid flies perform elaborate courtship dances, the red-eyed males tempting the females by swirling and twisting their wings in a manner reminiscent of a dancing geisha girl with her fan. Others are weirdly outlandish and mesmerising in their strangeness – for example, male stalk-eyed flies have their eyes on the end of very long stalks that protrude from opposite sides of their heads. Most flies never bother us to the slightest extent, living out their diverse lives beneath the human radar, and most of those 240,000 species have not been studied at all. For example, the larvae of most of the European species of dance fly have never been seen, even though they are common insects. We have no idea what they look like, where they live or what they eat – the adults just appear as if from nowhere every summer. What fascinating discoveries must await future dipterists.*

A small number of species are more noxious and have given flies a bad name that most do not deserve. Some bite, which is irritating; try having a picnic in western Scotland in August and you will be consumed alive by midges. Some of those that bite spread diseases – mosquitoes, blackfly, tsetse flies, and so on. Much as I love biodiversity, I'm pleased to say that, apart from the odd mosquito and horse fly, we have few biting flies at Chez Nauche. However, there is one fly species that sometimes appears in abundance, and of which I am not fond.

I have probably given the impression that Chez Nauche is a rural idyll, paradise on Earth, and in my eyes this is not a million miles from the truth. However, there is one small creature that does its level best to spoil the peaceful atmosphere – one literal fly in the ointment – and that is the house fly. Periodically during the summer months plagues of house flies arrive, having bred in piles of manure on nearby farms. They gather in swarms around the picnic table and around the doors and windows, finding their way inside before long. My wife Lara and I try to keep the doors and windows shut, and from the beams we hang dozens of fly papers, which quickly become covered in a revolting mass of struggling, sticky flies. We arm the children with fly swatters and set them to work, but their most enthusiastic efforts seemingly make no dent in the multitude. Like the Spartans at Thermopylae, the bodies mount up in front of them, but countless more flies pour in through gaps in the door frame and under the old clay

---

* Technically flies belong to the insect order Diptera, meaning 'two wings', differentiating them from most flying insects, which have four wings.

tiles in the roof, and eventually the boys are overwhelmed and give up.

This is not the first time I have fought a futile and losing battle against house-fly invasions. In 2006 I inherited a PhD student named Jason Chapman from a retiring academic. Jason is from Cardiff, a proud and vocal supporter of all things Welsh, and one of a select group of people I have come across who is truly passionate about flies. Jason's PhD was on trying to develop novel, environmentally friendly means of controlling house flies. The house fly, properly known as *Musca domestica*, is a greyish-black, smallish, bristly fly that is found wherever in the world there are people, animals and their dung, which is to say virtually everywhere these days. They have been around for perhaps sixty-five million years in something close to their current form, but really began to thrive as human populations grew and spread around the world. House flies probably evolved in the Middle East, but from there they have followed people around the globe. They can survive in the coldest climates by living in cattle sheds and chicken houses, helped by the warmth created from lots of animal bodies and fermenting dung.

The adult flies lay their eggs in any damp, rotting vegetable matter where there will be plentiful bacteria for the maggots to eat. Since the dung of most herbivores consists of damp, half-digested plant material mixed with bacteria, this is perfect. The maggots grow quickly and, once fully grown, crawl away to a drier spot to pupate. Under ideal conditions the whole life cycle takes about fourteen days, and a single female can lay 500 eggs in her life. It is easy, if a little pointless, to calculate the potential for population growth. One female fly leads to 500 flies in two weeks, 125,000 flies in four weeks, thirty-one million flies in six weeks, seven billion flies in eight weeks, and so on. Thankfully this never actually happens, as cold weather, outbreaks of disease or attacks from predators tend to curb population growth eventually, but nonetheless this explains how very large populations of flies can quickly appear when the conditions are favourable.

Of course rotting vegetable matter isn't very nice, and neither is animal dung, so something that thrives on eating it ought perhaps to be encouraged, welcomed, even admired. Dung beetles perform a similar function and were so valued by the ancient Egyptians that they carved huge stone statues of them. The problem with flies is not the maggots that eat the dung, but the adults. They have the most intensely annoying and deeply unhygienic habits. Their mouthparts are rather like a sponge on a stalk, with which they mop up more or less any soft, damp, digestible organic matter. If the food is too dry they vomit on it, puddle the vomit around a little with their mouthparts and then suck up the resultant mess. Their favourite comestibles seem to belong to one of two categories – animal faeces and human food. If they can't find either of those, flies will happily land on animals or people, preferring the lips and the corners of the eyes, or any cut or running sore, where they mop up any juices they can find. Failing that, they will happily drink a little sweat, vomiting occasionally as they go. The persistence with which they repeatedly land on ourselves and on our food is incredibly vexatious*. They have lightning reactions,

---

* 'Vexatious' is not a word one reads every day. When I read in an end-of-year report from one of my PhD students, who shall remain unnamed, that house flies were vexatious, I recalled that I had read the very same phrase in Jason's thesis a year or two earlier. In fact, when I checked, much of the report had been lifted word-for-word from the introduction of Jason's thesis. The unnamed student got into a lot of trouble, not least because – when questioned – he didn't even know what 'vexatious' meant, hesitantly suggesting that it might mean to fly in circles. Plagiarism has become one of the banes of academia, for it is all too easy for lazy students to patch together essays from Internet sources.

making them hard to swat, and they return over and over again, as if taunting us. A swarm of flies can turn a relaxing picnic into a nightmare.

If only they were simply annoying, it wouldn't be quite so bad. The real problem is that their behaviour makes them the most incredibly efficient vectors of an appalling range of diseases. Their stomach juices are chock-full of bacteria and viruses from previous meals on faeces, rotting food, dead animals, and so on. So when they vomit on my Saint Agur,* they are contaminating it with all manner of potentially dangerous disease organisms. Amongst other diseases, house flies are known to spread cholera, typhoid, salmonella, polio, hepatitis, *E. coli*, tuberculosis, anthrax, parasitic worms and conjunctivitis. More than 100 human diseases are spread by the little fiends, including many that can be fatal.

Now I strongly object to the modern obsession with sterility. Antibacterial sprays and hand-wipes may be vital in hospitals, but in my view they have no place in domestic homes. I'd much rather my children ate food with a bit of dirt on it, and grubbed around in ditches and ponds, than that they were kept in a sterile bubble. But I'd really prefer that they do not catch cholera, so flies go too far, even for me.

To return to Jason's PhD, his remit was specifically focused on trying to find a way to control house flies on battery chicken farms. These are among the most unpleasant places that man has thus far managed to create. If you've never been in one, count yourself lucky; continue to avoid the experience, and stop

---

* A magnificent soft blue cheese – if you haven't already tried it, you really should – but, sadly, also very popular with flies. Disappointingly there is no such place as Saint Agur, for I would have liked to go there on a cheese-related pilgrimage.

buying cheap battery-farm eggs. Battery farms are vast, dusty, foetid places. The chickens sit in tiny cages ranked three or four high, one above another. The floors are mesh, so that the excreta from the topmost chickens dribble down on to those below – if ever you are reincarnated as a battery-farm chicken, pray you don't end up on the bottom row. The floor of the building is slatted, so that the combined excretions of all the chickens eventually make their way through the gaps in the slats into a vast chamber beneath the floor, which slowly fills with rotting chicken faeces. This is bulldozed out every few months when it is full. The whole building is heated, perhaps because many of the chickens tend to be more or less bald and would die otherwise, but it is kept in sepulchral gloom to save electricity. The stench inside a chicken house is initially horrendous, but disturbingly it goes away if you are unfortunate enough to spend a long time inside.

You will have worked out by now that I am not a fan of battery farms. I speak from long experience, as my first paid job as a teenager was as a weekend egg-collector on the local battery farm. Modern farms have conveyor belts to automatically gather the eggs as they are laid, but in those days they were all collected by hand. It is a very odd feeling being the only person in a huge dimly lit shed with several thousand chickens all staring at you – of course they don't have much else to look at, since your appearance is the only remotely interesting thing that has happened to them all day. The job involved pushing a trolley loaded with egg cartons along each aisle and scooping up the eggs from both sides. It was almost too dark to see the eggs along the bottom row, so the easiest method was just to run your fingers blindly along the rack, picking up the eggs by touch, but occasionally you would accidentally grab the head of a dead chicken, which had slumped down on to the bottom of its cage.

Almost worse, some of the eggs were deformed, having no shells or having strange bulges. The deformed eggs were placed in a special bucket and were sent off to make shampoo. Thanks to these early experiences, I was not especially delighted to take over supervising Jason and renew my acquaintance with battery farms.

The problem he was trying to solve is easily explained, but exceedingly difficult to address. The accumulating mass of chicken faeces under the floor provides a perfect environment for house flies to breed. The fermenting mound remains warm throughout the year, and there is a constant supply of new food falling down from above. Flies can breed in their millions in these conditions. The adult flies are drawn upwards by the dim light filtering down through the slats above, and so they gather in countless numbers in the chamber above with the chickens. As if their life wasn't miserable enough, the poor chickens have to put up with the constant bother of flies landing on them. The flies can rapidly spread diseases amongst the chickens, and they make life intolerable for the humans who have to venture inside. Jason's job was to find a way of controlling these flies.

You might think this would be easy enough. There are plenty of chemical insecticides on the market – cans of fly spray can be bought from any supermarket. There are two problems with this approach. First, many insecticides are pretty unpleasant chemicals that are also toxic in varying degrees to both chickens and people. Some types of insecticide were originally developed as agents of chemical warfare – nerve gases – during the Second World War. Many have been banned, particularly in situations where they could contaminate human food – such as eggs – and so the selection available to a farmer is limited. The second problem is that flies have evolved resistance to more or less all of the insecticides that are available. Agrochemical companies are doing their best to develop new ones, but it takes many years and investment of millions of pounds to do so, so new insecticides reach the market very infrequently.

Of course there are those who deny that evolution takes place, but the development of resistance in flies to insecticides provides one of the neatest and most incontrovertible examples of evolutionary change, happening so rapidly that we can watch it in action. The speed of evolution depends upon the size of the population, the length of their generations and the strength of the selection pressure. Large populations with short generation times can evolve very rapidly in response to a strong pressure such as regular exposure to a deadly insecticide. Imagine a population of one million flies living in a chicken shed. The farmer sprays them with the latest type of insecticide. Just as in any population, the flies are all slightly different. They vary in size, wing length, the thickness of their cuticle, their colour, and so on. In part this variation is due to their genes, the combinations of which vary from individual to individual. Most will be carrying a small number of recent mutations, genes that were altered by a mistake during copying. Because of this variation, just a tiny proportion of them – let's say one in 1,000 – have a gene that makes them a little more resistant to the chemical than their fellows. Perhaps the gene allows them to break down the insecticide in their body, or perhaps it makes their cuticle a little less permeable to it. The vast majority of the flies are killed by the insecticide, but most of the few that survive carry one of the formerly rare genes that confer resistance. Freed from competition, these few flies breed rapidly in the abundant dung, and soon the population springs back towards one million. The farmer sprays again, but this time most of the flies are resistant, and his chemical has far less effect. Within a few months the chemical achieves nothing at all. The flies have evolved.

In reality, the process usually takes a little longer – resistance is rarely absolute, and it may take the spread of several genes, each of which confers a little more resistance, while at the same time the farmer applies heavier and heavier doses of insecticide to try and achieve a good rate of control. But in the end the outcome is more or less inevitable.

Of course the evolution of resistance is not confined to flies. In the tropics, many insect pests have developed resistance to insecticides. Insect pests breed much faster in warm climates, and there is no winter to check their growth. There are some horror stories from the middle of the twentieth century in which pesticides were grossly misused and many human lives were lost as a result. The cotton farmers of Central America and the southern states of the USA provide an example. In warm climates cotton suffers from a huge range of insect pests, including various moth caterpillars. When insecticides became widely available in the late 1940s and 1950s, cotton yields went through the roof for a few short years, as the insect pests were all but wiped out. Inevitably and inexorably, the pests started to creep back, and farmers responded by applying higher and higher doses of insecticide. When this ceased to work, the farmers would switch insecticide, and for a while all would be well, until the pests became resistant to the second and then the third chemical. Farmers then resorted to mixing them together into toxic cocktails, and applying them more frequently and in ever-increasing doses; some cotton crops were being sprayed sixty times per year. Eventually even this didn't work, and the farmers reached a point where they were spending so much on chemicals that they were no longer making a profit. They were also exposing themselves and their labourers to dangerous levels of some very unpleasant compounds, leading to all sorts of chronic health problems in rural communities, and to hundreds of deaths. To make matters even worse, there was no easy way back. The years of chemical use had wiped out all of the natural enemies: the predatory insects and birds that, in the days before insecticides, had helped to keep pest populations under control. The predators tended to have smaller populations and longer generation times than the pests they ate, so they had not generally developed resistance to the chemicals.

To return once more to Jason's flies, clearly blitzing the chicken houses with insecticides was not going to be the answer. His intention was to develop a 'lure and kill' system, whereby the flies would be attracted to a particular place, where they could be killed or trapped. There is a commercially available pheromone for house flies, a chemical known as z-9-tricosene, which is supposed to attract them, and Jason experimented with mixing this with insecticide and sugar, then painting the sticky mix on to boards that were hung from the ceiling of the chicken houses. You might question why he was using insecticides after all that I have told you about resistance, but their use in baited lures such as this seems to be much less likely to result in the evolution of resistance – flies that land on the boards consume so much pesticide that even those that have some resistance tend to die, so the resistance genes do not spread, or at least not as rapidly. Using pesticides in this way, rather than as aerial sprays, is also much better, as the amount of pesticide required is much less; and the pesticide is less likely to contaminate the chickens, their food or the farm labourers.

These boards did work – Jason would sweep up piles of flies from under each one, and count them – but there was a fundamental flaw. Most of the corpses in these mounds were males, which seemed to be preferentially attracted to the pheromone. One male fly can mate with dozens of females, so reducing the male population has a negligible impact on the size of the next generation of flies. What was really needed was something that attracted females too. Jason tried adding black spots to the boards, either evenly spaced or in clusters. Flies seem to be attracted to black spots, perhaps because they mimic flies, suggesting that there may be food or mates. In fact male flies seem to be attracted to any small dark object that is roughly the size of a female fly – they will even pounce on, and try to mate with, knots tied in a black shoelace, although they soon realise their mistake. Jason found that spotty boards worked a little bit better than boards without spots, producing slightly larger piles of dead flies beneath them, but they were still mostly males.

Just as he was coming towards the end of his PhD, I was watching the local news one evening when there was a piece about a fly outbreak on a landfill site near Portsmouth. The landfill was in close proximity to a smart marina, with expensive waterfront flats, posh restaurants, designer boutiques and moorings for very expensive yachts. A vast horde of flies had left the landfill site and infested the marina, showing a particular and understandable attraction to the restaurants, which presumably offered tastier delicacies than they were used to. The scale of the problem was such that the ceiling of one restaurant was black with flies, and of course they had to close until the problem could be solved. The residents and business owners were up in arms, threatening to sue the landfill company for compensation for their lost earnings.

A bit of digging the next day revealed the name and telephone number of the company. Rather speculatively I rang them up and offered my services, suggesting that they were clearly in need of specialist fly advice, which I could provide. They enthusiastically accepted my offer and invited Jason and me to visit the site. I had never before visited a landfill, and it was an eye-opener. The scale of the operation was impressive. I can't remember the exact figure, but the site was taking thousands of tonnes of refuse every day, with a constant stream of lorries bringing it from all over Hampshire. Huge bulldozer-like machines spread and compacted the waste, which, as you can imagine, consisted of a horrendous mix of plastic bags, paper, soiled nappies, rotting food, broken toys and more or less anything else you can think of. The toys seemed particularly poignant – once-loved dolls and cuddly toys, dog-eared and moth-eaten, doomed to be buried in a sea of filth. Immense numbers of gulls and crows wheeled above the heavy machinery, swooping down and bickering over any tasty morsels that were revealed. Others, presumably replete, perched on the surrounding fences or sat amongst the rubbish. A strong sea breeze was coming in from the south-west and noisily shredded the loose flaps of plastic shopping bags and bin-liners, but thankfully it also whipped away the terrible sour stench. Oddly enough, Jason seemed to like the smell, and said it made him feel hungry.

At first sight, it seemed to be no great surprise that the landfill site was a source of large numbers of flies. People clearly throw away a lot of organic matter: kitchen scraps, chicken carcasses, and so on. Together with the used nappies, there was an awful lot of food for maggots. The only surprise to me was that there hadn't been outbreaks of flies before – the site had been in operation for a couple of years.

The staff explained how the site worked. They would excavate huge pits, line them with thick black plastic sheeting, like giant garden ponds, then fill them with waste. The traditional system of managing landfill was to periodically cover the latest layer of waste with a compacted layer of soil, which kept down the smell and stopped rubbish from blowing away. However, landfill space is in short supply, and taking up valuable space with soil was expensive, so an alternative approach had recently been adopted:

instead of covering the waste with soil, a huge sheet of sacking would be rolled out over it at night, then rolled back in the morning to allow more waste to be added. The company had also recently switched to fortnightly rather than weekly waste collections. The staff speculated that the fly outbreak might have been linked to one of these changes, and we agreed to investigate. We also set up a fly-outbreak early-warning system, consisting of dozens of sticky traps on the landfill site and at 'control' sites away from the landfill. Every week the traps would be collected and the number and species of flies carefully counted. It seemed likely that fly outbreaks might be stimulated in part by suitable weather conditions, and so we obtained weather data from the local Met Office and built mathematical models to see if we could predict how many flies there were likely to be the following week. If we could predict a major outbreak, then perhaps action could be taken to control it before it became too awful – although we weren't sure what that action might be. At the very least the landfill company could get its lawyers on standby for another rush of complaints.

We conducted an experiment to find out whether the move to fortnightly waste collections had contributed to the problem. Any organic waste in rubbish bins is likely to attract female flies looking for somewhere to lay eggs. If it is carefully bagged, or in bins with tight-fitting lids, then flies ought not to be able to get to it, but inevitably bags get ripped and bins get left open. When bins are full to overflowing it is impossible to shut the lid. The longer the waste is in the bin, the worse it is likely to smell. Thus it seemed likely that the switch to fortnightly waste collections meant there would be more maggots in the rubbish when it was collected, but we decided we had better verify that this was indeed so.

Setting up the experiment was an awful job. In some areas refuse was still being collected weekly, and the landfill company directed us to a waste depot where we could obtain samples of refuse that were from either weekly or fortnightly collections. We shovelled up a tonne or so of each and sealed it in large plastic rubbish bins. Each bin we fitted with an 'emergence trap' at the top: a clear plastic bottle into which any emerging adult flies would crawl. Every day for the following few months we collected up, counted and identified the flies emerging from the two types of rubbish. It was surprising just how many species of fly emerged from the waste – we found more than fifty different species, including bluebottles and greenbottles, which feed on rotting flesh; various small midges and gnats; and of course many house flies. Predictably, more flies emerged from the waste collected after two weeks. What was perhaps more important was that the flies emerging from the two-week-old waste started emerging much earlier, some of them almost as soon as we set up the experiment. This was not surprising – house flies have a two-week life cycle in the warm summer months, so if waste was contaminated with fly eggs as soon as it was placed in the bins, then by the time the waste was collected two weeks later the eggs would have had time to hatch and the maggots to grow to full size and pupate.

To see whether the switch from using soil to cover the waste to using sheets of sacking was likely to have made the company's fly problems worse, we carried out some simple experiments to establish how good flies were at burrowing out of the ground when they emerged from their pupae. We buried pupae under a layer of loose soil or compacted clay, and also sealed them under samples of sacking. When a fly first crawls from its pupa its cuticle is soft, and it has an expandable balloon in the middle of its face, which it uses as a hydraulic jack to burrow upwards. It is a rather freakish sight. The fly pumps liquid into the balloon and then deflates it, creating a small space. It then crawls forward into the space and repeats the process. Flies were pretty good at burrowing up through soft soil, but could not get through clay or sacking. However, since the sacking was rolled back every morning, it simply served as a blanket to keep any flies that emerged during the night warm until they were released. It was pretty clear that the old system of using a layer of clay had sealed in many of the flies in the waste, and that the switch to a sacking blanket was making life much easier for them.

We told the landfill company that their problems were largely down to the fortnightly collection and the new sacking system, but they were extremely reluctant to change. They were saving an awful lot of money – so much so that even the threat of being sued by the nearby restaurant owners was not sufficient to force a rethink. They wanted us to find an alternative solution to their fly problem.

We tried many approaches. We spent a long time analysing the smell of male and female flies, in the hope of finding a pheromone that would attract females. If such a compound could be found, perhaps it could be used on insecticide-and-sugar-painted boards, as Jason had done in the battery farms, but placed outdoors around the landfill sites. We even synthesised a blend of chemicals which contained almost all of the compounds found on a real fly – 'fly in a bottle'. Some of these chemicals and blends did attract flies in the lab, but whenever we tried them outdoors on landfill sites, they were more or less useless. The windy conditions, the vast numbers of real flies and the background stink of the landfill itself seemed to mask any attractant effects of our chemicals.

The only aspect of our work that was really any use at all was our fly-monitoring system. Our network of sticky traps around the landfill site gave us a pretty good early-warning system. When it was combined with weather data, we were able to predict forth-coming fly outbreaks with moderate accuracy – warm, humid weather usually preceded an outbreak, while heavy rain seemed to reduce their numbers. When we warned the landfill company of an impending outbreak, they reverted to drenching the site in insecticides. Although not 100 per cent effective, and certainly not very environmentally friendly, it usually seemed to damp down the numbers of flies for a while.

As we battled for a solution, the landfill filled up with waste. Eventually the vast hole that had been excavated was full, but still the waste arrived. Such is the shortage of landfill sites that permission is often given to create hills out of the waste, and so it was here. The waste mound slowly grew and was sculpted into a rounded hillock. We were no nearer a solution to the fly problem when the landfill site was finally closed. The waste was sealed over with a plastic membrane, with pipes buried in it to allow methane gas to escape, so as to avoid the whole lot exploding. The plastic was covered in soil, then sown with grass seed and planted with shrubs. Today all that can be seen is a slightly unnatural-looking grassy knoll just to the south of the M27 near Portsmouth.

As soon as the waste stopped arriving the fly problem ceased, and the nearby residents and restaurateurs must have breathed a huge sigh of relief. Now the waste goes somewhere else – there are thousands of landfill sites in Britain. Most, sensibly, are not adjacent to towns or fancy marinas, for there is still no really effective way of controlling the flies they produce.

That landfill sites exist at all is testimony to our staggering short-sightedness. 'Sustainability' is a word that is bandied about a lot these days, with good reason, for if we selfishly use up the Earth's limited resources while polluting and contaminating it with our waste, then our children will have to manage without, while having to clear up our mess. Burying our waste in holes in the ground, or creating hills out of it, is clearly not a sustainable practice. The Earth is, increasingly, a small and crowded planet. There are only so many holes, and most are already full. Former landfill sites are of little use for anything much, for inevitably the waste slowly subsides, and one day the plastic membranes that seal in the festering waste will split.

Where will our children put their waste? In Japan they have taken to building landfill sites out to sea: walling off shallow areas, pumping out the water and then filling them with waste. How long before an earthquake or storm ruptures these, spilling millions of tonnes of decaying waste into the sea?

The solution is obvious. We should recycle everything. It is perfectly possible. An awful lot of the packaging that we throw away was entirely unnecessary in the first place. All packaging could be made from recyclable material. If all organic waste were composted, there would be nothing for flies to feed on in landfill sites; and if everything was recycled, there would be no more landfill sites.

There is a certain irony that the pests we try to control often resist our best efforts, while the creatures that we do not wish to harm, such as bees and butterflies, often become the unintended victims of our clumsy attempts. Whatever we do, there will always be house flies. They may be hard to love, but they are one of nature's success stories: adaptable, prolific, invincible. They will be here long after we are gone, when there is no one to be vexed by their habits. So long as there is something rotten and smelly somewhere, flies will continue to efficiently convert it into more flies, and to provide food for swallows, lizards and praying mantises. They may not be very likeable, but no doubt they will continue to swarm occasionally at Chez Nauche, whether or not I enjoy their company. In a way, I take satisfaction from the fact that there are some creatures we cannot bully into submission;

that, for all our intelligence and technology, we are unable to make more than the slightest temporary dent in their numbers. Good luck to those filthy flies!

CHAPTER SIX

The Secret Life of the Meadow Brown

24 May 2009. Run: 38 mins 20 secs. People: one ancient man tending his splendid allotment, weeding his long rows of carrots and beans. Dogs: 6. Butterfly species: 13. A horseshoe bat flew in an open window last night, but then couldn't find the way out, so it spent half the night circling above my bed, wafting me with its wings before I was forced to get up and catch it with my butterfly net and release it outside – hence I'm feeling a bit groggy this morning. On the last leg of my run I spied a ruderal bumblebee queen nectaring on the yellow rattle in the meadow – a wonderful, huge beast, a zeppelin of the insect world, with chocolate and velvet-black stripes. I also saw my first meadow brown butterfly of the year, a male, no doubt the first of very many more to come.

One of the most common butterflies to be found in my meadow at Chez Nauche is the meadow brown, properly known as *Maniola jurtina*. If you get down on your haunches on a sunny day in July and scan across the waist-deep vegetation of the meadow, you will see hundreds of them, flitting about from flower to flower and occasionally indulging in aerial dogfights. Indeed, meadow browns are amongst the most common butterflies to be found in more or less any meadow in Europe south of the Arctic Circle. They fly for ten weeks or so, from early June through July and much of August, but spend most of the year as green, hairless caterpillars quietly munching away on grass. They are decidedly drab, with a milk-chocolate upperside to their wings, and a camouflaged underside marked with wavy fawn and greyish bands. They generally sit with their wings closed, so that only the undersides are visible, and when they do so they are very hard to see amongst the yellowing, sun-bleached grasses of a summer meadow. If you look very closely you may see between one and five very tiny black spots along the margins of the hindwing, sometimes with a faint white halo surrounding them. These spots don't look like much, but I spent two and a half years of my life counting them, and trying to make sense of the numbers I obtained.

To explain why, I need to take you back to Oxford in the 1940s, and to the eccentric career of Edmund Brisco 'Henry' Ford. The 1940s was an odd but exciting time for biology. Darwin's concept of evolution by natural selection had revolutionised biological thinking, but the modern science of genetics did not yet exist. It was entirely unclear how evolution worked at a mechanistic level. Gregor Mendel, an Augustinian friar working in Austria in the 1860s, became posthumously famous for his experiments with peas, which demonstrated that characters were inherited in discrete units of some unknown material, which were passed from parent to offspring. We now use the word 'gene' to describe these units, and we know that they are made of DNA (deoxyribonucleic acid), but Mendel had no way of knowing this. He worked tirelessly for many years, rearing more than 29,000 individual pea plants from crosses that he had made by hand. One single experiment took him ten years. Sadly, the significance of his work was unrecognised at the time; if only his contemporary, Charles Darwin, had come across it, he would surely have realised that here was the very mechanism that underpinned inheritance, and hence his theory of evolution. Mendel's work was rediscovered twenty years after his death in the early 1900s, but the heritable material was not identified as DNA until the 1940s, and even then no one knew how it carried information. It was not until James Watson, Francis Crick and Rosalind Franklin elucidated the structure of DNA in 1953 that genetics could really take off as a discipline, and it is staggering to think of the progress that has been made in the sixty years since then, with entire genomes now being sequenced.

To return to Ford, he was born in 1901 and spent his whole career based at Oxford. By the 1940s he was a lecturer in the Zoology Department. He more or less single-handedly founded a discipline that became known as 'ecological genetics' – the study of genetic change in natural populations – at a time when genetic change was impossible to measure directly, since no one knew what genes were made of. These days, modern genetic tools allow us to sequence genes and compare sequences between individuals and populations, but none of this was available to the early eco-logical geneticists. Instead Ford and his colleagues were forced to use visible differences between individuals, and between species, that might (or might not) serve as a proxy for genetic differences. Butterflies and moths had long been known to show variation in wing patterns between individuals; indeed, a popular and (with hindsight) rather eccentric hobby in the early twentieth century was to collect unusual variants of particular butterfly species and pin rows and rows of them in cabinets. Ford had collected butter-flies as a youth and so he was aware of this variation.

Ford was convinced that natural selection was all-important in nature. At around the same time other geneticists, most notably the American Sewall Wright based at the University of Chicago, were developing theories which suggested that chance events alone could also lead to genetic change in populations. With a simple pen-and-paper argument, Wright showed beyond reasonable doubt that gene frequencies could change rapidly in small populations, due to a process that eventually became known as 'genetic drift'.

Wright's work has long since been accepted, and indeed it forms the basis of our understanding of inbreeding in rare animals, which can and do rapidly lose genetic diversity through drift. Nonetheless, Ford seems to have taken offence at this notion, and spent much of his career trying to prove that Wright was wrong.

Ford looked for evidence for natural selection in the changing frequencies of different colour morphs of moths and butterflies. He settled on a trio of study species: the peppered moth, the scarlet tiger moth and, of course, the meadow brown. Peppered moths, the best known of the three species, exist as either a very pretty pale form, which has white wings decorated with an intricate pattern of black-and-grey spots, or as an entirely black form. The pale form is beautifully camouflaged when sitting on the lichen-crusted trunk of a tree. Naturalists had previously noted some remarkable changes in the relative abundance of the pale and black forms. In 1811 the black form was rare, but by the 1860s it was more common than the pale form in urban areas, and by the 1890s almost all peppered moths in cities were black. In the middle of the twentieth century the change began to go into reverse, with the black form becoming less common, and now it is quite scarce once again. Ford had a PhD student, Bernard Kettlewell, who studied this phenomenon and provided the experimental evidence to explain what was happening. In fact it was all fairly simple, and to this day provides one of the neatest textbook examples of natural selection in action. The genetic basis of the colour difference was easily established by lab crosses – the wing colour is controlled by a single gene, which gives the moths either the pale or black form. So the frequencies of the different forms of the gene had swung markedly over 100 or so years, but why? It turns out that it was all to do with pollution. Lichens are very susceptible to pollution, and during the Industrial Revolution they largely died off in cities. Indeed, the soot produced by early industry caked the trunks of trees so that they were more or less entirely black. Kettlewell used what are known as 'mark-release-recapture experiments', whereby moths are caught, marked with a dot of ink, released and then as many as possible are recaptured a few days later. In the blackened woodlands of industrial Birmingham, he found that the pale moths were much less likely to survive to be recaptured than the dark ones, but when he repeated this experiment in an unpolluted, lichen-rich woodland in Dorset the reverse occurred. He released both moths and great tits into large aviaries and found that, depending on the backgrounds available for the moths to perch upon, the birds tended to find and eat whichever colour morph most poorly matched the background. Pale moths stick out like a sore thumb to birds and humans alike, when resting on a soot-blackened tree, and so during the Industrial Revolution the dark moths had a huge advantage and more or less replaced the pale ones. When cities were cleaned up in the early twentieth century, the pale form started to creep back.

It is perhaps a shame that Ford didn't stick to studying peppered moths. The two other species that he chose to study produced much more equivocal results. The scarlet tiger moth is a very pretty red-and-black moth with cream-and-orange spots. It exists in three different forms, the most common of which has many spots, plus a rarer form with some spots missing, and a very rare form that is largely black. The moth itself is rather rare, being found near Oxford in only a handful of small sites. Ford studied one site in particular, Cothill, an odd little patch of fenland of just a couple of hectares, surrounded by woodland. This little patch of soggy ground seems to suit the moth, perhaps simply because there is plenty of its favourite food plant, comfrey, and so tiger moths are plentiful in most years. The adults are rather sedentary, and in the daytime in June and July they are easily spotted, sitting around on the vegetation. Ford counted the numbers of the three morphs each year from 1939 more or less until his death in 1988. He found that the frequency changed rapidly from year to year – so rapidly that it could surely only be explained by natural selection of some sort. Drift can't result in rapid genetic change in large populations over one or two generations, as appeared to be the case here. Unfortunately it was never clear what the selection was – scarlet tiger moths are brightly coloured to advertise the fact that they are poisonous, so it was unlikely to be anything to do with predation. Nonetheless, Ford used his data from Cothill to castigate Sewall Wright in a series of articles from the late 1940s to the 1970s. With hindsight it all seems rather odd, for Wright never tried to claim that natural selection didn't occur – he simply argued that chance events could also be important. Hence, even if Ford's data had been watertight, it couldn't possibly have proved Wright wrong.

And so we return to Ford's third species, the meadow brown. But before we do so, let me explain how I became interested in it. In 1987 I graduated from Oxford with a degree in biology and the notion that I was going to become a conservationist and save the world. But first I decided to cycle across the Sahara with an old school chum (another Dave). We duly did so, more or less, ending up somewhere near the border between Algeria and Niger before flying home just before Christmas – more tanned, a little thinner and a lot poorer than we had left. We both signed on the dole, and joined the local Shropshire Conservation Volunteers to keep ourselves busy. A couple of months later I spotted an advert for a PhD on the ecology of butterflies in Bernwood Forest, based at Oxford Brookes University. I had pretty much ruled out doing a PhD, because I didn't think I was cut out for academia. I also didn't much fancy another three years of scrimping and saving on a student stipend (these days PhD stipends are more generous). But this PhD offered what seemed at the time like a reasonable salary in exchange for some basic teaching duties. What was more, I had always loved butterflies, and by bizarre coincidence I had done my undergraduate final-year project on the ecology of butterflies in this very same forest, which lies about eleven kilometres east of Oxford. It seemed like fate, so I applied and, sure enough, I got the post.

I started my PhD in April 1988. My PhD supervisor, Denis Owen, was an odd character. He had spent many years working out in Africa, and would often come to work in colourful East African gowns. He smoked like a chimney in his office, so much so that the room filled with smoke and he became a blurred, dimly visible silhouette when viewed through the glass window of his office door. He would spend his time in there hammering away on an old-fashioned manual typewriter so loudly that the thin office walls would gently shake. Denis barely spoke to me when I started, merely telling me to go and spend a lot of time in the field, looking at butterflies. I was rather intimidated and so didn't push him for any more pointers, although I was in desperate need of them. I hadn't really got a clue what to do. A PhD requires an idea, a hypothesis, an interesting question to pursue. I had none of these, so I went every day to Bernwood and wandered about looking for butterflies.

Bernwood has an interesting history. One thousand years ago it was a vast royal hunting forest, covering many hundreds of square kilometres, and it is said to have been a favourite hunting ground of Edward the Confessor. Over time it was gradually cleared and converted to arable land and pasture, as successive monarchs sold off chunks of the forest to raise cash. By the early twentieth century Bernwood had shrunk to occupy just 300 hectares, but despite its diminished size it had become well known as a hunting ground for butterfly-collectors. It supported many rare woodland species, including the spectacular purple emperor and the black hairstreak. Sadly, it was purchased by the Forestry Commission in the 1950s, and at the time this organisation had little interest in butterflies or in conservation. Virtually the entire wood was clear-felled and sprayed with herbicides from the air to kill off any regeneration. Dense stands of conifers were planted, and then sprayed with DDT to kill beetle pests that attacked the conifers. Remarkably, the forest remained a hotspot for rare butterflies through the 1960s and 1970s. Despite the herbicide sprays, woodland flowers and shrubs somehow survived, and so did most of the butterflies. When I first visited Bernwood in the mid-1980s, it still had many rare butterflies; it remains the only place in the UK where I have seen wood whites, brown hairstreaks and purple emperors. It was thought to have more breeding butterfly species than any other site in Britain. Sadly, though, most were in decline. The conifers were growing larger and larger, casting a dense shadow and thus shading out all of the native flora that provide butterflies with nectar and food plants for their caterpillars. My undergraduate project was on the effects of increasing shade on woodland butterflies, and its conclusion was that all but the most shade-tolerant butterflies, such as the speckled wood butterfly, would soon disappear.

Luckily, the Forestry Commission has evolved over time. Where once it was concerned solely with efficient timber production and with generating maximum profit, now it is far more interested in conserving biodiversity and managing forests in a way that balances commercial needs with those of wildlife. Changes were evident in the late 1980s at Bernwood, where the Forestry Commission began widening rides and cutting small clearings to combat the gloom cast by the towering spruce-tree plantations. It was pretty small-scale stuff, but it was a start.

I spent the spring and summer of 1988 walking these rides, looking for butterflies, and trying to come up with an idea as to what I should study for my PhD. Having a strong interest in conservation, I decided to study the ecology of a rare species, the Duke of Burgundy fritillary. It isn't actually a fritillary, and it has no known connection to the Duke of Burgundy, so its name is hard to fathom, but it is a lovely little creature with a chequered orange-and-brown upperside and silver spots on the underside of its wings. The butterfly is on the wing in May, while for much of the rest of the year the caterpillars can be found on the leaves of their food plants, cowslips and primroses. At least in theory they can. In practice I walked through Bernwood every day in May when it wasn't raining and saw one Duke of Burgundy, a female that promptly flew away. I searched diligently for eggs and caterpillars over the following months, but found none, although I did map out the locations of every primrose in Bernwood. By September I had exactly one data point, and there really wasn't a great deal I could say based on that.

I spent the autumn and winter reading everything I could on butterflies, in search of inspiration. It felt as if time was slipping away. My PhD funding was for three years and there are, self-evidently, just three field seasons in three years; I had squandered one already. If my second field season was similarly unsuccessful I would be doomed to fail my PhD. It dawned on me, not before time, that studying rare creatures is rather difficult. For safety, I decided instead to study something very common, something that I could not fail to find by the thousand. And this is where we return to the meadow brown. It isn't a woodland butterfly, but there were quite a few in the woodland rides and I had seen countless numbers of them in an adjacent meadow, known appropriately enough as Bernwood Meadows. But what exactly should I study? I came across Ford's work on meadow browns and became intrigued. I knew of Ford, for he was author of the New Naturalist book *Butterflies*, which I had been given as a child. I'd also met him briefly when I was an undergraduate student at Oxford, at which time he had been a very old but nonetheless rather intimidating figure.

Ford worked closely with Wilfrid Hogarth 'Bunny' Dowdeswell, a former Oxford undergraduate who became an academic at Winchester College, and together they counted the spots on the hindwings of meadow brown butterflies for twenty-five years, from the late 1940s to the 1970s. They obtained samples from all over Europe, and particularly from the Scilly Isles, which they visited repeatedly over many summers, camping together on the uninhabited islands.* In most populations in Britain and Europe, the majority of male meadow browns have two spots, while most females have none. In some peripheral populations, such as some of the Isles of Scilly, females with one or two spots were very common. On some of the smaller of these islands the frequency of spotted and unspotted butterflies varied sharply from year to year, far more than could be explained by chance (or genetic drift, as Wright termed it). This, Ford declared once again, was clear evidence for natural selection, and proof that drift was inconsequential. What is more, Ford described a mysterious boundary phenomenon along a line roughly following the border between

---

* It seems very likely that Ford was gay. He lived at a time when homosexuality was illegal and subject to horrendous persecution. He never married, had no children and was a prominent campaigner for the legalisation of homosexuality. He seems to have held women in low regard. He campaigned vociferously against their admission to Oxford, as either undergraduates or fellows. A story passed on to me as an undergraduate at Oxford was that on one occasion he turned up to teach a class to find that only female students were present. Legend has it that he gazed around the room as if looking for someone and then announced, 'Since no one is here, today's class is cancelled.'

Devon and Cornwall, where the frequency of spotted and unspotted butterflies changed markedly from one side of a hedge to the other. Given that meadow brown butterflies can readily fly over hedges, this seemed odd, but was again attributed to some powerful selective force at play.

What interested me was whether these spots actually served any function that could explain the selective forces that seemed to change the frequency of more or less spotty individuals from place to place and year to year. Some butterflies, such as the peacock, have very prominent eye-spots near the tips of their forewings, which are thought to scare away potential predators by mimicking the eyes of some much larger beast. The eye-spots of a peacock are hidden when it sits with its wings closed, but if it is disturbed, it has only to open its wings to reveal huge eyes comprising concentric circles of blue, red and yellow. Owl butterflies – enormous creatures that live in the rainforests of South America – have eye-spots and associated wing patterns that convincingly resemble the face of an owl. Meadow browns also have far less impressive eye-spots near the tips of the forewing, on both the upper and lower sides, formed from a large black spot with a small white spot in the centre, resembling the glint of light reflected from the black pupil of an eye. These spots might serve to frighten predators, but this wasn't especially interesting to either Ford or me, as all meadow browns were the same in this respect.

Could the small spots on the underside of the hindwing have a similar purpose, relating to predation? Many hairstreak butterflies have an elaborate false eye near the edge of their hindwing, often with slender tails that resemble antennae, creating the impression that the back of the insect is in fact the front. To further enhance this impression, hairsteaks often perform a swift 180-degree turn as soon as they land on a perch or flower. The idea is that this may deflect the peck of a predatory bird from the real head of the insect to the margin of the wing, which readily snaps off, allowing the butterfly to escape. It is common to find butterflies with peck-marks nipped out of the edges of their wings.

In the early 1980s Paul Brakefield, an English scientist based in Leiden, proposed an explanation for the variation in spot numbers in meadow browns and their relatives. He suggested that the spots did serve to deflect attacks, but that the trade-off was that they might make the butterfly more obvious to predators when sitting still. When flying about, visiting flowers and looking for mates, butterflies are very obvious to predators such as birds, so deflective spots might be helpful. In contrast, if the spots make them less well camouflaged when stationary, they would be best not to show them when sitting still. Female butterflies spend much more time sitting still than males, who fly around a lot, which might explain why females tended to have no spots whilst the majority of males had two. Brakefield went on to predict that in populations inhabiting cold, wet climates, where the butterflies spend a lot of time sitting around waiting for the sun to come out, spots ought to be a disadvantage and hence rare, whilst in sunnier places the spots ought to be more common. The latter certainly appears to be true in the large heath butterfly, a relative of the meadow brown in which spottiness decreases in populations that are further north or west or at higher altitudes.

This theory was quite neat, but no one had ever shown that these spots actually did deflect the attacks of predators. I decided to try to find out. I also became interested in the genitalia of meadow browns. The genitalia of male butterflies consist of a pair of rather barbaric claspers, which lock on to the female during mating, and between which the tube-like phallus protrudes. Place your wrists together with the palms of your hands facing one another and you get a rough idea of the design of the claspers. They are hinged at the base, so that they can open out and then close on the female like grappling irons. In some butterflies the claspers are armed with sharp, incurved, talon-like hooks to help them grip. Pity the female speckled wood butterfly. In the meadow brown there are no sharp hooks, but there is a bristly 'thumb' protruding from their upper edge. The genitalia of male butterflies, and of male insects in general, tend to be very useful in identifying species that are otherwise similar. They usually differ in obvious ways between species, whilst being fairly uniform within species. One theory that had been put forward to explain this was that the male and female genitalia are like a key and lock; they only fit together if the male and female are of the same species, helping to prevent unfruitful mating between members of different species. The exception is the meadow brown, in which the male claspers are very variable. In some the 'thumb' is no more than a gentle bump, while in others it is very pronounced with a swollen end capped in bristles, resembling a tiny pollarded willow tree. So far as I could tell, no one had ever looked at the fit between female and male insect genitalia before, or explained why meadow browns have such variable genitalia, and this seemed worth exploring.

Questioning the function of tiny spots on butterfly wings, or explaining why their genitalia come in various shapes and sizes, might seem fantastically obscure subjects to pursue. Answering these questions was never likely to change the world and, with the benefit of hindsight and experience, I can think of many more profound topics that I might have tackled, but in the spring of 1989 this was all I had.

To get to grips with the function of wing-spots, I counted their number in meadow brown populations all over southern England. I had a lovely blue 650cc Suzuki motorbike on which I roared about the countryside, with a net, notebook and countless small cardboard pots in my rucksack. I wanted to test whether butterflies had fewer spots in cooler, shadier sites and more on sunny, south-facing sites, as Brakefield's theory would predict, but disappointingly this didn't seem to be true. I went down to Cornwall to look for Ford's boundary, where the proportion of spotted butterflies was said to change suddenly, but it was no longer there, so far as I could tell. I didn't seem to be getting very far, so I tried more experimental approaches.

I killed a range of spotted and unspotted butterflies, dried them with their wings open or closed, and then tied them to grass seedheads in Bernwood Meadows and waited to see whether they were eaten by birds. Butterflies with their wings open were more likely to be scoffed, but whether or not they had spots on their hindwings appeared not to make the slightest difference.

I tried a different tack. I caught 600 butterflies from Bernwood Meadows and changed their spots artificially. It is quite easy to remove these small spots by simply brushing away the dark wing scales with the tip of a damp paintbrush, and equally it is easy to add small black spots with a marker pen. So I randomly assigned individuals to either spotted or unspotted groups, made the necessary changes, gave them an additional discreet mark on the topside of their wings, so that I could subsequently identify them, and released them back where I had found them. Three days later I went back and hunted for my marked butterflies, to see which ones were still alive. In theory, if females spend most of their time sitting still, spots should make them more conspicuous and hence more likely to be eaten, while males that spend most of their time flying around should benefit from having spots to deflect attacks. So I predicted that females with added spots and males with spots removed should be predated more highly that unspotted females and spotty males. Of my 600 butterflies I managed to recover ninety-one – it was hard going because there are countless thousands of meadow browns in that meadow, so finding the ones I had marked was like looking for a needle in a haystack. To my slight surprise, the results followed the prediction. In particular, adding a spot to females did seem to make them more likely to be eaten.

Studying the genitalia proved much harder. To measure them the poor butterfly has to be killed, partially dissolved in strong alkali, and then the genitalia pulled out and spread on a slide to look at them under a microscope. Studying them in action was still more challenging. There were umpteen published studies describing the genitalia of male insects of different species, but no one seemed to have looked to see what exactly they did when in contact with the female. Did they really form a key that perfectly matched the female lock? Finding mating butterflies in the field is fairly easy, particularly with common species such as the meadow brown. They can sometimes be seen flying around, the larger female carrying the male dangling behind her, but usually when mating they sit still on a stout grass stem. However, at the slightest disturbance they tend to separate. In any case it is impossible to see what is really going on with the naked eye, as the genitalia are too small. I needed somehow to persuade the mating couple to sit on the stage of a microscope, but that was hardly practical.

After much thought I came up with a cunning plan. I filled a square polystyrene cool-box with liquid nitrogen, strapped it to the rear seat of my bike and rode out to Bernwood. Liquid nitrogen is very cool stuff, in both senses of the word. It boils at -196°C. It can render rubber or metal so cold that they become as fragile as glass – even the most expensive padlock will shatter like a cheap plastic toy if dipped in liquid nitrogen and then given a gentle tap with a hammer. When it is exposed it rapidly boils, releasing super-cool nitrogen gas, which freezes water vapour and carbon dioxide in the surrounding air.

As I banked around the roundabout at the top of the Headington Road, some liquid nitrogen spilled, creating a brief cloud of white vapour behind me, which must have momentarily confused other motorists. But otherwise all went well. Once in the meadow I searched for mating butterflies, carrying the cool-box with me. Before long I spotted a pair, dangling from one of the seedheads of a tussock of cocksfoot grass. I took the lid off the cool-box, placed the container beneath them, then gave the seedhead a firm tap. As I had hoped, the pair tumbled down into the liquid nitrogen and were instantly frozen, still locked together in copulation. I left them there and collected a few more in the same fashion, before returning to the lab at Oxford Brookes.

Back at the university I set up a microscope in one of the walk-in freezer rooms, and there I could examine my frozen couples at leisure. It became clear that only parts of the male valves were actually in contact with the female. To return to my analogy of paired hands, the main area that gripped the female was the lower edge – the little finger and side of the palm. The 'thumb' was nowhere near contacting the female, so presumably it didn't matter what shape it was. This seemed to explain why this part of the valve was so variable: if it doesn't really do anything, then natural selection would not act upon it, whereas those parts that grip the female would be under strong selective pressure to match precisely the female's shape.

I investigated all of this further with a second experiment in which I tried to measure the strength of the bond between male and female. I wondered if any attributes of the male genitalia – such as size or the shape of the 'thumb' – might affect the strength of the bond. If I was correct in thinking that the thumb wasn't gripping the female, then there should be no relationship. I thawed out my couples, clipped the female to a stand and then attached successively heavier weights to the male, to see at what point the pair fell apart. Of course it might not have been hugely realistic, since the butterflies were dead and so presumably mating with less enthusiasm than usual, but it was the best I could come up with. It worked and, as predicted, the size or shape of the thumb did not seem to affect the strength of the bond. I never did explain why male meadow browns have more variable genitalia than other butterflies. It would have been good to study how the genitalia of lots of other species fit together when mating, but I didn't have time.

During my PhD I also studied whether spottier butterflies tended to fly more than unspotty butterflies – they did not. I looked to see if spottiness affected how often a butterfly mated, or whether it affected mate choices – it did not. I spent a long time grinding up butterflies and running their enzymes along electrophoretic gels in the lab, to get an indirect measure of how much meadow brown butterflies move about in the landscape. The answer was: quite a lot. I tested how long butterflies could sustain flight at different temperatures, and whether they differed in the lower temperature limit at which they could fly, but it didn't reveal much of great interest – more- or less-spotty butterflies all performed much the same. I heated pupae gently from one side, and found that this produced adults with more and bigger spots on the warmer side. This rather undermined much of the work that I had done on spots, and Ford and Dowdeswell before me, for if they were influenced strongly by the environment, then differences in spottiness between populations could simply be due to micro-climatic differences. The whole point of Ford's work was to study genetic change, on the assumption that the visible differences in wing patterns of individual butterflies reflected genetic differences. If they simply reflected whether the caterpillar had happened to pupate in a slightly warmer or cooler spot, then the whole thing was a bit of a waste of time.

I followed this up by doing a little work on whether temperature affected the wing patterns of scarlet tiger moths. Oddly, Ford never seems to have bred scarlet tigers himself, to prove beyond doubt that the different colour forms were genetically controlled, although they are quite easy to breed in captivity. I collected some moths from Cothill and did some simple rearing experiments at different temperatures. I found that storing pupae at slightly elevated temperatures resulted in adults that had lost some spots and resembled Ford's intermediate colour morph. If switches between colour morphs could be brought about by changes in temperature, then all Ford was recording throughout those decades may simply have been evidence that the weather at Cothill varied from year to year. I suspect that Sewall Wright would smile, if he were still alive.

Looking back, it is fair to say that my PhD was not a fabulous success, but fortunately I passed. My supervisor didn't really speak to me at all until I was writing it up at the end of three years, and at the time I slightly resented that he gave me such scant advice, but with hindsight I perhaps learned more than I would otherwise have done by having to think for myself. Coming up with lots of duff ideas and trying them out was frustrating, but I guess it may have helped me work out what good ideas look like. That said, I do try to give my own PhD students rather more help than I received.

When next time you are lucky enough to find yourself in a meadow in high summer, keep an eye out for the meadow brown. They are not glamorous as butterflies go, but they have an understated charm. The decades of intense scientific scrutiny are long forgotten, and they seem content to go quietly and unfussily about their unspectacular business, the tiny black dots on their wings mattering not one jot. It seems remarkable now that they could ever have been the subject of such intense and heated debate.

Butterflies are one of the most beautiful of insect groups, for the wings of many types are stunningly colourful, or tastefully ornate, or subtly but intricately shaded, and all possess a symmetry that is lovely to behold. We know that some patterns seem to provide camouflage, that big spots might frighten predators and that bright colours might please a mate or warn a predator that the butterfly is poisonous, but beyond that we can hardly begin to guess what purposes their beauty serves – if, indeed, it serves any at all.

CHAPTER SEVEN

Paper Wasps and Drifting Bees

5 July 2009. Run: 37 mins 27 secs. Personal best! I was perhaps helped by a vicious little black spaniel in the hamlet of L'Âge Marenche, which chased me, nipping at my heels for 100 metres or more. People: none. Dogs: 7. Butterfly species: 18, including a fabulous Queen of Spain fritillary, with silver spots flashing on her underside as she flew. I also got a distant view of a pair of stone curlews in a fallow field to the north – peculiar boggle-eyed, long-legged birds that prefer running to flying. They are a great rarity, even here.

Karl Marx was right, socialism works, it is just that he had the wrong species.

E. O. Wilson

Between the farmhouse at Chez Nauche and the northern boundary I have planted an apple orchard. I'm not really sure why, but I've always loved orchards. My grandparents on my mother's side had a lovely orchard at the end of their large garden in Norfolk. My grandfather had planted the trees himself as a young man – not the short, stunted trees on dwarfing rootstocks that are always used these days (it makes the fruits easier to pick), but proper trees that had grown ten or more metres high. Picking the fruit was a hazardous business requiring long wooden ladders, and much of it fell to the floor, where it was fought over by my grandfather's free-range chickens and swarms of wasps. It was a lovely place in late summer, alive with the buzz of insects. (Sadly, it is now a housing estate.) Old orchards with full-sized trees are rare these days, and are havens for insect and bird life, so I really wanted one in France, but there was a problem. I couldn't find anyone who sold apple trees on rootstocks that would allow them to grow to full size. In any case I couldn't really afford to buy enough of them for a proper orchard, so I decided to grow my own from pips. I deliberately bought a range of apple varieties and collected the pips from every one I ate. I sowed them in the greenhouses at Southampton University, pretending that they were for an important research project. They grew well, and after a year were big enough to plant out. My dad often came down to France with me and it was he who planted them out in 2004 and 2005. He planted fifty in total: five wonky rows of ten, with a generous spacing of ten metres between each tree. The ground is pretty stony and hard to dig, so he often had to use a pickaxe to make a hole, but he stuck at it stubbornly and got there in the end. Each small tree he protected against rabbits with a green plastic tube just over a metre tall, tied to a wooden stake.

To my father's continuing annoyance, cows break into the meadow from the farm to the south on a regular basis, and invariably make a beeline for the orchard. The stakes seem to make perfect scratching posts for them, but more often than not the weight of the cow leaning against it snaps the stake and pushes the tree over to a precarious angle, so that the orchard needs endless running repairs. My dad seems to have taken this maintenance as a personal battle between himself and his bovine arch-enemies.

Every year since the orchard was planted, when I go down to Chez Nauche for my first spring visit, I rush eagerly to see how the apple trees are coming along. Apples don't self-pollinate, so the pips will all grow into crosses between the variety of apple that I ate and whatever it was that pollinated it. This has made watching the trees grow particularly exciting, as I have little idea what type of apple each tree will one day produce. It was not until 2011 that I got my first blossom, on one of the most northerly trees near the hedge. Some bees must have brought pollen from apples in the village a kilometre or so to the east, for the blossom duly set and by late summer the little tree was carrying dozens of tiny, bright-red crab apples. They looked beautiful, but were toe-curlingly sour. Crab apples are great for jelly and cider, but I do hope some of the trees one day produce apples that I can eat.

For the first couple of years the trees weren't tall enough to poke out of the top of the tube, and I had to peer down to see how they were doing. This, I soon discovered, was a hairy business, because almost every tube contained not just a small apple tree, but also a paper-wasp nest. The tubes seem to provide a perfect, warm and sheltered environment for these wasps, which zoom out of the top of the tube aggressively if disturbed. However, they probably eat any greenfly or other pests on the apples, and don't seem to do any harm, so they are welcome to their tubular homes as far as I am concerned.

I first came across paper wasps in 1995, when teaching on a field course in the south of Spain. Every year for the eleven years that I was at Southampton University we took all of the first-year biology students for a fortnight to a beautiful corner of Andalucia, about eighty kilometres west of Gibraltar, on the Atlantic coast. It was an epic undertaking, for there were usually more than 100 students, but it was great fun and a wonderful way to enthuse them with the excitement of field biology. The area is extraordinarily rich in wildlife. Rugged, craggy mountains cloaked in cork oaks rise steeply from the sandy beaches and coastal dunes, and griffon vultures soar about the steep cliffs. We went in late March, when the wild flowers are in riotous bloom in the rough meadows, rocky slopes and the damper slacks behind the dunes, and they attract innumerable bees, hoverflies and butterflies, such as the spectacular red, black and white Spanish festoon. This is also the time when birds that have migrated to Africa for the winter return to Europe, and many choose to cross here from Morocco, which is clearly visible across the chilly blue-grey Straits of Gibraltar. Storks, black kites and bee-eaters arrive in large flocks, often looking tired and flying low to the water, for there are no thermals to lift them when crossing the sea. The area is also particularly rich in reptiles: turtles bask on the banks of the streams, viperine snakes hunt the tadpoles of toads and tree frogs in the shallows, skinks, geckos and eyed lizards scamper amongst the sun-warmed rocks, and chameleons perch motionless in the bushes.

Many of the farmers in the area used prickly-pear hedges to separate one rough field from the next. Prickly pears are horrible plants. They are a native of the Americas, and have become very troublesome invasive weeds in other parts of the world, most notably in Australia. They have numerous exceedingly sharp spines and, to make matters worse, the spines are covered in microscopic barbs. This makes them very hard to pull out, once embedded in skin, and they often snap off and then cause festering sores. Having said all this, the prickly pears in Spain do provide excellent refuges for lizards and snakes, which can bask with impunity amongst the spines in full view of the frequent booted eagles that would otherwise love to eat them. They also provide a very popular nest site for paper wasps.

Paper wasps are, of course, not made of paper. They resemble the familiar yellow-and-black social wasps that plague picnic tables in late summer, but they are more slender and delicate in appearance. Their name comes from the material from which they construct their nests, which is, essentially, home-made paper. However, since many other wasps also use paper for their nests, the name isn't especially helpful. Paper wasps and the common 'picnic-table' wasps all make nest material by chewing at dead dried wood and then regurgitating the chewed-wood fragments with saliva, to make a greyish paper from which they can construct the most elaborate combs of hexagonal cells. Garden fence panels and posts and untreated garden sheds all seem to be much favoured by wasps as a source of wood for paper, and they can often be seen – and even heard – chewing away at garden woodwork. The nests of social wasps in the UK can grow to a very large size, sometimes bigger than a basketball, but paper-wasp nests are, like the wasps themselves, relatively small and delicate. They have a single stalk supporting a plate of up to 100 or so hexagonal cells in which the offspring are reared. They make no protective shell around the nest, so that the growing grubs (and their tending by the adult wasps) are easily observed.

We would spend the first week of the field course in Spain teaching the students to identify plants and insects, and then in the second week they would conduct small research projects on a topic of their choice. It was always a bit of a challenge trying to find enough different projects for 100 students, and the paper wasps struck me as a species that could be interesting and fairly easy to study. On my first trip to Spain in 1995 I had not seen paper wasps before, and I spent a little time watching the nests. The wasps seemed to engage in a lot of fighting; they would regularly face one another, eyeball to eyeball on the surface of the nest, and then push and bite until one or other retreated. At the time I didn't know anything about paper wasps, but this seemed intriguing. What were they fighting over? Did the same wasp always win? I got a couple of students to do a project that aimed to find out. They marked the wasps with different-coloured paints so that they could recognise individuals, and then watched the nests all day long.

It did seem that fights had predictable outcomes – one large female seemed to be the boss, and she would shove, bully and harass the other females on the nest. What was even odder was that new, unmarked wasps, which had not been seen before, seemed to turn up regularly on the nests, even though it was clear no pupae had recently hatched. Some of the marked wasps also disappeared, but we presumed they had died. It was when the students reported that a marked wasp from a different nest had arrived and, seemingly, integrated without much trouble into her new nest that I became really interested. I had only just begun to work seriously on bumblebees at the time and had not studied other social insects in any great detail, but I had never heard of social insects moving between nests. As an undergraduate I had been taught how ant and bee colonies consisted of a queen and her daughters, and how the close relatedness within the nest made this social system possible. I knew that ant colonies would often fight ferociously, and that honeybees would try to steal honey from each other's nests, and would post guards on the entrance to the hive to keep out intruders. It didn't seem to make any sense for wasps to move around like this, or for them to be welcomed into their new homes.

I became very excited, and pushed the students to mark huge numbers of wasps on every nest they could find. They got rather prickled by the pears, but stuck at it. They found that movement between nests was quite common. Some individuals seemed to move back and forth, switching their allegiance from day to day. To me this seemed most peculiar, and I excitedly planned writing up this discovery for *Nature*, a top-notch journal.

When I returned to Southampton I was in for a big disappointment. I dug out the literature on paper wasps and soon found that their movement between nests had already been described. Nonetheless, it was fascinating to read about them, as their biology was so similar to in some ways, but markedly different in others, to that of bumblebees. The social structure of bumblebees is seemingly simple. Each nest has a single queen and many daughter workers. Many ants and social wasps are much the same. Paper wasps, on the other hand, are decidedly flexible in their nesting arrangements. Nest-founding begins in spring when the females emerge from hibernation. Some females try to found a nest of their own, as does a bumblebee queen. Other females do so working as a small group – an uneasy alliance – although usually there is a hierarchy, with one female being clearly dominant. These groups often, but not always, consist of sisters, thought to recognise one another by smell and perhaps also by facial recognition: paper wasps tend to have highly variable face patterns. The nest produces female workers, most of them the daughters of the dominant foundress, and grows steadily through the spring and summer, producing new foundresses and males in late summer, much as a typical bumblebee nest does. However, unlike bumblebees, the workers are capable of mating, and sometimes do so. If the foundress dies and there is no subordinate foundress, then a worker will mate and take over her role. Both subordinate foundresses and workers sometimes switch allegiance from one nest to another, perhaps tending to move to nests containing related individuals. If a nest is destroyed, as often happens due to attacks from predatory birds or mammals, then the foundresses and workers that survive will usually join another nest, or sometimes try to kill the foundress on another nest and take it over. If they succeed in the latter, they kill and eat the young brood in the nest so that they can replace it with their own, but they allow the older grubs and pupae to develop, presumably to increase their workforce. This makes single-handedly founding a nest a high-risk strategy for a female paper wasp, for her nest is easily commandeered by others. In all of this each wasp is presumably trying to maximise her reproductive success, either by helping relatives to reproduce or by producing her own offspring, wherever possible.

Paper-wasp societies seem to consist of a network of shifting, fickle alliances, rather than the more rigidly organised nests of bumblebees. Or so it seemed to me in 1995, but this turned out to be incorrect. Since then we have been slowly discovering that bumblebee nests are not quite what they seem. This has only become possible through advances in genetic 'fingerprinting' techniques, which enable us to identify who is related to whom. Of course these techniques are not only useful for bees; they have also been used in studies of many other animals, including humans, and the results are often surprising. For example, swallows were long considered to be monogamous. These charming, elegant birds form into pairs, which work together to build a nest and rear a clutch of young. I regularly have two or three pairs nesting in the rafters of my old barns at Chez Nauche, their cheerful and excited twittering as they gather mud for their nest from the edge of rain puddles in the yard a sure sign that spring is well under way. It was only when the DNA of the offspring and parents was compared that it became apparent that pairs of swallows were not quite as devoted to one another as they appeared. Many nests contained chicks that were the offspring of the mother, but were unrelated to their apparent father. The females clearly sneak off to engage in illicit copulations, so their cuckolded partners may spend much of their time rearing someone else's offspring, while also presumably trying to obtain a few extra mating opportunities themselves.

This relates not just to swallows; chronic, serial infidelity seems to be the norm in most birds. Great reed warbler females routinely sneak off to mate with males other than their partner, preferring trysts with males that have an impressive song repertoire. And some birds take infidelity to extremes. Males of the superb fairy wrens – beautiful little Australian birds with turquoise-and-black markings – spend much of the breeding season sneaking off to court females in nearby nests, even resorting to picking flowers to present to them in the hope of being rewarded with a swift sexual encounter in the bushes. Of course it is not just birds that are unfaithful. Genetic studies of various human populations have revealed that anywhere between 2 and 30 per cent of children are the result of what biologists term 'extra-pair copulations'.

In birds, such studies sometimes also reveal that offspring are unrelated to either parent. This seems to be particularly common in many species of duck, which routinely try to lay their eggs in the nests of others if they get the chance, hoping that their offspring will be looked after by somebody else. This most frequently happens in birds that have large clutches of eggs, perhaps because an extra egg or two in a nest that already contains a dozen eggs is not very obvious to the resident female when she returns. Of course cuckoos take this strategy to its logical conclusion, entirely abandoning building their own nests in favour of laying their eggs in the nests of other species.

It turns out that bumblebees are rather like ducks, at least in this respect. Worker bees are physiologically incapable of mating, but they can lay unfertilised eggs that develop as males. In 2004 bumblebee researcher Carlos Lopez-Vaamonde at London Zoo discovered that their experimental buff-tailed bumblebee nests were being infiltrated by unrelated workers that snuck in to lay eggs, essentially parasitising the nests of others. Some of these bees originated from other experimental colonies, but others were coming from wild buff-tailed nests in Regent's Park. If they succeed, these workers get to produce sons that they will not have to look after at all, thereby passing on some extra genes to the next generation and so improving their fitness in evolutionary terms.

This all seemed pretty fascinating, and I discussed it with Steph

O'Connor, who was working with me at the time as the handler of Toby, our bumblebee sniffer dog.* Interesting though the work from the zoo was, it seemed a bit unnatural. The nests being infiltrated were reared in captivity and were contained in artificial boxes connected by tubes to the outside world. It seemed to us that natural, wild nests might be harder to find and invade. Steph's project involved finding bumblebee nests with Toby, and then following how well they fared and what predators and parasites attacked them. She had found quite a few wild buff-tailed bumblebee nests in and around the university campus, and was planning to dig them up towards the end of the season when they had produced whatever new queens and males they were going to, in order to reveal what parasites were inside. It seemed like a great chance also to find out how many of the male bees produced by wild nests were the offspring of sneaky unrelated workers.

Steph duly dug up as many of the nests as she could. It was a horrendously difficult job, for some of the nests were at the end of tunnels three metres long, often down old rabbit burrows. The tunnels would frequently go under the roots of trees and deep underground, so that some nests were impossible to reach. The remaining bees weren't too impressed, either. Buff-tails can be understandably aggressive when their nest is threatened and, even though the nests were old and declining, some of the workers would fly out and attack. Steph would catch them, one by one, and place them in pots. She persevered and eventually excavated fourteen nests in their entirety: worker bees, queen, pupae, grubs and eggs. She then genotyped them all, which was a hefty job, since there were well over a thousand individual bees at one stage

---

* Toby was trained by the army to sniff out bumblebee nests, which are usually hidden underground or in dense thickets. Toby and Steph's exploits are described in *A Sting in the Tale*.

or another. After several months in the lab she was finally able to piece together who was related to whom. The results contained some surprises.

Most of the males in the nests were sons of the queen, so long as the queen was still alive. Some (usually less than 1 per cent) were sons of the nest's workers, and grandsons of the queen. This wasn't particularly surprising, since we had known for many years that workers try to lay their own male eggs, preferring to rear sons rather than brothers, if they can get away with it (although if their mother, the queen, catches them, she roughs them up and gobbles up the eggs that would have become her grandsons). In total four males were entirely unrelated to the rest of the occupants – proof that workers really do sneak into the nests of others to lay eggs under natural conditions. Six of the worker bees in the nests were not daughters of the queen – presumably these were sneaky bees from nearby nests that we had caught in the act; they might just have been nipping in to lay their eggs when Steph came along, or they might have permanently moved into a new nest, perhaps because their own nest was destroyed by badgers or disease.

All of this was as we might have guessed. The surprise was that some nests also contained groups of females, including both adult workers and eggs, that were sisters to one another, but entirely unrelated to the rest of the nest. A group of adult sister bees could all be sneaky workers coming from a nearby nest, but clearly eggs do not move from nest to nest on their own, and it seems highly unlikely that adult bees would carry them. These bees had to be the result of a second queen, who had somehow snuck in and laid a batch of eggs.

In bumblebees 'usurping', whereby a queen invades a nest and attempts to kill the resident queen, has long been known and seems quite common. However, this was always thought to happen early in the spring when nests are small. Queens emerging late from hibernation might struggle to find a vacant nest site, but might instead find a young nest and opt to try to take it over.* Steph's nests were dug up in August and September, right at the end of the season. The nests had not been usurped; the resident queen was still in occupation in most of them. It appears that, just as workers opportunistically sneak into nests to lay eggs, so do queens. This seems rather odd, for it begs the question where these queens have come from. Perhaps they are old queens whose own nest has been destroyed somehow, and who are therefore forced to dash into the nests of others to lay batches of eggs in a desperate, last-ditch strategy to leave behind a few offspring. Or they might conceivably be new queens, just emerged from another nest and newly mated, but this seems unlikely, as the conventional wisdom is that queens do not mature their ovaries until after hibernation. A queen who poured energy into developing her ovaries in late summer would use up her fat stores and would probably be unlikely to survive the winter, so such a strategy would make little sense.

One of Steph's nests did contain a second queen, unrelated to the rest of the occupants. She had not laid any eggs, or if she had they had been eaten by the nest's residents. It is possible that she had attempted to usurp the nest earlier in the year, but had failed and

---

* 2013 was a bumper year for nest-usurping. We placed about 100 buff-tailed bumblebee nests out in the countryside in May, part of an experiment to see how many pesticides they are exposed to. It had been a very cold and miserable April, and perhaps this prevented many buff-tailed queens from founding their own nests. Whatever the reason, we found that almost all of our experimental nests were invaded by wild queens, with some accumulating as many as seven dead queens lying in the bottom of the nest, in addition to the resident. We don't know how many of these invasions were successful, and what proportion of the original queens survived.

then stayed on in the nest as a subordinate queen – rather as several mated females live together among paper wasps. Alternatively she may have arrived recently and been biding her time, in the hope of getting the chance to lay some eggs, or perhaps even attack and kill the resident queen, at some point in the future. Whatever the truth, studies such as Steph's are revealing that the lives of bumblebees are neither as simple nor as regimented as was once supposed. Nests do not just contain a queen and her offspring, but may be a mix of multiple queens, workers from a range of nests, sons and grandsons of the queen and adoptee sons foisted on the nest by sneaky workers.

Social insects have sometimes been held up as examples of ideal, altruistic societies, where all struggle selflessly for the common good. Luminaries as diverse as Aristotle, Virgil and Shakespeare extolled their virtues; Socrates even went so far as to suggest that the most virtuous humans might hope to be reincarnated as ants or bees. In reality, bee, ant and wasp societies are far more interesting than the utopian ideals for which they were mistaken. Ruthless power struggles that would put the Borgias to shame are commonplace, with murder and even cannibalism being frequent. There is little here that we might wish to emulate, yet there is still a huge amount that we might learn from studying these wonderful, fascinating creatures.

CHAPTER EIGHT

The Mating Habits of the Death-Watch Beetle

25 July 2010. Run: 39 mins 19 secs. Bit slow today, slight calf strain. I now creep quietly past the spaniel's house. People: none. Dogs: 2. Butterfly species: 21. There was a young fox on the drive; it obviously hadn't seen a runner before – its curiosity got the better of it, and it let me get very close. I also spotted a purple emperor butterfly sitting in a sunspot on the ground in the oak woods near L'Âge Marenche. What a beauty, its wings flashing in the sunlight; how I would have liked to add one of those to my collection when I was a boy! Optimistically I used to try the time-honoured method of placing a rotting rat on woodland rides – this was supposed to attract them to feed on the exuding juices – but it never worked, perhaps because the nearest population was 100 miles away from Shropshire, where I lived at the time.

Tom lay awake and waited, in restless impatience. When it seemed to him that it must be nearly daylight, he heard the clock strike ten! This was despair . . . he lay still, and stared up into the dark. Everything was dismally still. By and by, out of the stillness, little, scarcely perceptible noises began to emphasize themselves. The ticking of the clock began to bring itself into notice. Old beams began to crack mysteriously. The stairs creaked faintly. Evidently spirits were abroad. A measured, muffled snore issued from Aunt Polly's chamber. And now the tiresome chirping of a cricket that

no human ingenuity could locate, began. Next the ghastly ticking
of a death-watch in the wall at the bed's head made Tom shudder
– it meant that somebody's days were numbered.

Mark Twain, *The Adventures of Tom Sawyer* (1876)

The very first time I slept in the farmhouse at Chez Nauche I was in for a shock, and the renewal of an old acquaintance. It was June 2003, and the night had been warm, so I woke early. While summoning the energy to get up, I heard a faint tapping. Tap-tap-tap-tap-tap, pause. Tap-tap-tap-tap-tap, pause. I knew that sound. Its source was an animal with which I had once been very familiar, having spent six months of my life studying it. It was a creature that I should not have been pleased to hear – at least not inside my house – but I couldn't help but smile.

In early 1992 I was briefly unemployed, a dispiriting experience. I had finished my PhD on meadow brown butterflies the summer before, and had spent the autumn filling in as a replacement biology teacher at Eton College (an interesting time, but enough to convince me that schoolteaching was not my forte). I lived at the time in Didcot – someone had to – and going to the dole office in this slightly seedy town was depressing. I was madly applying for jobs all over the world, and I vividly remember being quizzed by an official as to my progress in finding work. She had a form to fill in, and one of the questions asked where I was looking for work. I explained that I had recently applied for jobs in the UK, Australia, the United States and Costa Rica, and suggested that she wrote down 'the world'. She frowned, unhappy, for the next question was 'Where else would you consider looking for work?'

Fortunately soon afterwards I heard about a short-term research post in the Zoology Department at Oxford University. When I was an undergraduate, I had spent a lot of time in the big, ugly, concrete zoology building on South Parks Road. The post was to work on the mating behaviour of the death-watch beetle, a creature about which I knew little at the time. Two zoologists, Martin Birch and Tristram Wyatt, had become interested in how death-watch beetles find their mates and had managed to obtain a small grant from the Association for the Study of Animal Behaviour. I applied, was interviewed by Martin and Tristram, and was delighted to be offered the job immediately.

Martin was once one of the leading experts in insect pheromones, but had tragically suffered brain damage in a car accident a few years earlier and was struggling to readjust. He had been a brilliant scientist and was still with-it enough to understand fully that parts of his brain weren't working properly, and would get enormously frustrated and angry with himself when he couldn't remember simple names or words. Nonetheless he was a lovely guy and great fun to work for.*

So what is so interesting about death-watch beetles? They are very small, a little over half a centimetre long, drab brown in colour and roughly bullet-shaped. They evolved to eat dead trees, and have since happily adapted to eating the timber beams in houses, which is lucky for them because in our modern, tidy world dead trees are never left lying around for long. Death-watch beetles grow incredibly slowly. From egg to adult takes up to thirteen years, reflecting the fact that there isn't a lot of nutrition to be extracted from a dead tree, and that burrowing through it is hard work. Given enough time (two or three centuries usually suffice), they can do a lot of structural damage to a building; the roof of Westminster Hall in London nearly collapsed due to a very prolonged infestation. Unless you are interested in the preservation of ancient buildings, these beasts don't sound terribly exciting. The clue to their interest lies in the name.

* Sadly Martin died in 2009, aged sixty-five.

In past times, before hospitals were common, most folk died at home. Imagine a timber-framed cottage, grandad lying in bed in his nightshirt slowly dying from the palsy or some other indeterminate (at the time) but fatal disease, the family gathered around his bed in respectful silence. A faint drumming can be heard coming from the walls, from the ceiling. Five or six beats in quick succession, like someone drumming their fingers, or the tip of a pencil being tapped on a wooden table. Legend had it that this faint sound was the Devil, impatiently drumming his fingers as he waited for his chance to snatch the soul when it departed from the body. The noise was rarely heard at other times, but only because most households were rarely quiet enough except during a 'death-watch'. Of course it was not the Devil – or at least not most of the time. As you might have guessed, it is actually the mating call of the death-watch beetle. These days few of us live in timber cottages, and those who do have almost certainly had the timbers treated against beetle infestations. The poor death-watch beetle has become a rare creature, so few of us have ever had the chance to hear its ardent call.

A death-watch beetle spends almost all of its long life as a small white grub, equipped with tough mandibles with which it chews its way very, very slowly through timber. After a decade or so it is fully grown, and in the autumn it forms a pupa. Pupae hatch into adult beetles the following spring. They live for just a few weeks as adults, and feed hardly at all. As with most insects, the adults exist to mate, and to lay their eggs. The beetles are slow-moving, clumsy and nearly blind. They have a poor sense of smell, so far as we can tell. Finding a mate presents something of a problem, to which they have evolved an unusual solution. They bang their head against the wood. Given their tiny size, they do so remarkably hard, making a noise loud enough for us to hear quite clearly, and sending vibrations through the surrounding timbers. Only the males bang their heads spontaneously. They pause in their wanderings, brace themselves with their forelegs and raise their heads, then swiftly strike their forehead five or six times within perhaps half a second, producing the characteristic drumming. If you place a death-watch beetle on a tin can it makes quite a racket.

The female makes exactly the same noise, but only in immediate response to a male, and only when she is a virgin. A solitary male drums occasionally as he walks. This was the noise I heard in Chez Nauche: Tap-tap-tap-tap-tap, with a long pause as the beetle ambled about, then tap-tap-tap-tap-tap once more. If a virgin female is about, the two beetles indulge in duets, drumming to each other. If several virgins are present, they will all reply to a male's drumming, at which point the male becomes really quite excited, running about and beating his head repeatedly against the wood.

Now this had all been known for many years. My job was to work out how they found each other. You might think this a very trivial question – they can hear each other, so of course they can find each other easily, can they not? Well, no. Martin and Tristram had realised that these tiny beetles could not locate the source of a sound in the way we do. When we hear a noise, our brain automatically compares the volume in our two ears and uses this to compute the rough location of the sound. It is not very accurate, but if the sound is repeated a few times, we can usually work out where it is coming from, more or less. Death-watch beetles hear with their feet, sensing the vibrations in the wood beneath them. Their feet are perhaps two or three millimetres apart, and so a vibration will not attenuate measurably between reaching one foot and the next. The beetles might obtain some directional information if they could detect the timing of the arrival of the noise at each foot, but quick calculations based on the speed of movement of vibrations through timber suggest that the vibrations would arrive at the first and last foot of the beetle just three-millionths of a second apart, which should be beyond the abilities of any nervous system to detect. So the replies of a female beetle tell a male that a virgin female is somewhere round about, but – frustratingly for him – they do not seem able to convey any information as to exactly where she is. Yet clearly death-watch beetles do mate, or else they would not still be with us. So do they just wander around and rely on blind luck to find each other (in which case isn't all the drumming a bit of a waste of time?), or is there more to it? It was my job to find out.

The first task was simply finding some beetles to study. As I have said, these are no longer common creatures. Forests nowadays tend to be tidy affairs – dead trees are quickly cut down and disposed of, either because they might harbour tree pests and diseases or because they might fall on passers-by. This, combined with pesticide treatment of old buildings, means there are few places left to find creatures that live in dead wood. Fortunately Martin knew of an old church in north Oxfordshire, in the picturesque village of Steeple Aston, which had a large population of death-watch beetles. Every couple of days we would drive up there and search the floor and pews for beetles that had fallen from the ceiling above. We found it was easiest to borrow the broom that was propped in a corner and sweep the dust-bunnies and beetles into a pile, then sort through it. The church was always cool and the beetles always very lethargic, so it was an easy job to gather them up. We could often find twenty or thirty beetles in one visit. Back at the lab, I would place each in an individual plastic pot and keep them all in a cool room until I was ready for them.

I managed to borrow a huge disc of solid hardwood, a slice of rainforest tree, from the Forestry Department. This gave me a large, flat wooden arena on which to study my beetles. To start with, I would take a virgin female beetle – as ascertained by her willingness to reply to male calls – and place her in the centre of the disc. I would then place a male at the edge and simply watch to see what happened. The answer was, usually, not very much. Either the male would simply sit there or wander off without drumming, showing no interest in finding a mate, or he would drum, but the female would decide that she wasn't in the mood and wouldn't reply. Trying to get both a male and a female interested in each other at the same time was proving tricky, so I hit upon the idea of building an electronic beetle that could play the role of either male or female. It didn't need to look like a beetle, just sound like one. After much fiddling about I ended up with a very crude gadget, the business end of which consisted of an old loudspeaker from a radio, at the centre of which I glued a plastic pipette tip. This was mounted in a clamp stand, so that the tip of the pipette touched my wooden arena. The remainder of the set-up was an untidy bundle of wires and electrical circuits which, at the touch of a button, would send a pulse of electrical signals to the speaker, which would turn them into sound, and this was transferred into vibrations in the wood by the pipette tip. After much adjustment I could mimic the drumming of a beetle and, what is more, if I so chose I could make my robotic beetle drum faster or slower, louder or more quietly, or with more or fewer beats than a real beetle.

I put the robotic beetle at the centre of the arena and began placing males on the edge. Whenever one drummed, I would reply immediately and, as far as I could tell, the beetles were completely fooled into thinking that the reply came from an eager virgin female. If the male was initially sleepy or inactive, a couple of drums from the robot usually stimulated it into life. In this way I could see how good the males were at finding the 'female'. It soon became clear that they were pretty hopeless. They would wander around, drumming their heads and listening for a reply, often taking hours to approach the location of the robot. I filmed their behaviour and spent many hours recording their tracks, in an attempt to discern any kind of pattern.

The answer turned out to be fairly straightforward and unexciting. When a male beetle is getting replies from a female, he tends to walk a short way and then drum his head again. He listens carefully to the reply. If it is louder than the previous one, he knows he is going in roughly the right direction, and so he keeps going, pausing at intervals to repeat his call. If he is going in the wrong direction, her reply will get quieter – when this happens, he tends to turn round and head back the way he came. If she stops replying entirely (suggesting that he has walked so far from her that she can no longer hear him), he tends to turn round and go back. By this very crude mechanism, he should on average get slowly closer to her. I wrote this up as a scientific article and sent it off – my job was done. But I still had a few weeks of funding left, and I had become intrigued by something else.

When a male finds a real, virgin female he invariably clambers straight on to her back and attempts to mate with her. The males are not very bright, so they often climb on to the female backwards (to be fair, the front and back ends of the female don't look much different). He encourages her to mate by continuing to drum, but now of course he is whacking his head against the back of the female's head or, if he is the wrong way round, against her backside – an interesting form of foreplay. At the same time he extrudes his genitalia and probes away hopefully. The interesting (but perhaps not entirely surprising) thing is that often this less-than-subtle approach is not successful. Mating can only happen if the female is willing, because she also has to extrude her genitalia from under her carapace. If she refuses to do so, there is nothing the male can do about it, but continue to drum away and prod at her with his penis. If this doesn't get him anywhere, after a few minutes he usually gives up and wanders off to find a more obliging female.

I noticed fairly quickly that the larger males were more likely to succeed in charming a female than the smaller, weedier ones. I took to weighing the males before and after mating, and discovered something that seemed quite astonishing to me. On average, during copulation the males were transferring 13 per cent of their body weight to the female, and up to 20 per cent in some extreme cases. Exactly what the material was that was being transferred I never found out, but presumably it was sperm and associated juices. This is roughly the equivalent of a human male producing fourteen litres of sperm in one go – not an entirely pleasant prospect. Large male beetles tended to produce more, so by choosing the heaviest males to mate with and rejecting the small ones, the females were ensuring that they received particularly large packets of sperm. Why might the females do this?

It is not uncommon in other insects for males to transfer nutrients to the females during mating, as well as sperm. In some creatures, such as hanging flies and scorpion flies, the male catches a prey item and uses it to persuade the female to have sex with him, a so-called nuptial gift. The larger and more impressive the prey item, the more likely she is to be impressed. In some crickets, the male impregnates the female with sperm during mating, but also glues a large, sticky ball of nutritious gloop to her bottom, which she promptly eats. In all of these cases the food helps the female to produce more or larger eggs, which with luck will be fertilised by the male, so the nutrients go to benefit his own offspring. This nuptial gift might be particularly important in creatures like death-watch beetles where the adults don't get to feed much, if at all, so the females have no other source of nutrition. In this situation it might be vital for the female to choose a male that could provide the most nutrition, since this would help her to produce the maximum possible number of offspring. Obviously female death-watch beetles get an excellent opportunity to judge the weight of a potential mate when he is sitting on top of her, whacking her in the back of the head.

I experimented with playing tricks on the females. I found that I could fool a female into mating with even the smallest and weediest male, simply by putting a bit of Blu-tack on his back. The Blu-tack made the male heavier and that was all it took, although the poor female then received a most disappointing package.

One interesting side-effect of the large nuptial gift produced by male death-watch beetles is that it makes them weigh substantially less than they did before, which in turn makes them much less attractive to females. Even the biggest male could rarely persuade a female to mate with him if he had mated before, however hard he tried. The only way I could make him appealing to females was once more to make him heavier with a blob of Blu-tack. And once again the females must have felt short-changed, because whatever was being transferred to the female, the male could produce very little of it when he was mating for the second time. So it seems that in natural situations most male death-watch beetles probably only ever get to mate once, and then they have shot their bolt – unable to seduce another female because they weigh too little.

Eight years after I discovered death-watch beetles at Chez Nauche they are still there. They are slowly chewing through the timbers in the living-room ceiling, but they are huge old timbers and my guess is that it will take them at least another hundred years before they do any serious damage. I could inject the timbers with insecticide, but I haven't the heart to do so. These beetles have probably been there since not long after the house was built, 150 or so years ago, so who am I to evict them? There are few enough places left for these intriguing little creatures to live out their slow-paced lives, and in any case I would miss the sound of the Devil impatiently tap-tap-tapping his fingers.

CHAPTER NINE

The True Bugs

5 April 2011. Run: 38 mins 26 secs. People: none. Dogs: 8. The sneaky spaniel was lying in wait, and nipped my ankle before I saw the blighter coming. Butterfly species: just 7, a poor haul today. However, the lack of butterflies was easily made up for by spotting my first golden oriole – what a spectacular bird, so colourful it seems like an escapee from a zoo. As I ran by I unwittingly flushed it out from some elm bushes on the side of a farm track, but it briefly settled in an oak further down, so I got a reasonable look at it. A male in breeding plumage, the size and shape of a large thrush, but golden-backed with a black mask – a wonderful creature.

> *It is amazing what a lot of insect life goes on under your nose when you have got it an inch from the earth. I suppose it goes on in any case, but if you are proceeding on your stomach, dragging your body along by your fingernails, entomology presents itself very forcibly as a thoroughly justified science.*
> Beryl Markham, *West with the Night* (1942)

Every time I arrive back at Chez Nauche after any time away the place is overgrown, the path to the front door impeded by waist-high vegetation, and one of the first jobs I have to tackle is hacking down the foliage. In late spring this invariably disturbs scores of pairs of firebugs, chunky, flattened red-and-black bugs, which spend much of their lives as adults locked in endless copulation. For reasons best known to themselves, they seem to regard the area near my door as a prime location for their sexual activities.

I use the term 'bug' advisedly. It is of course a word that is widely used to describe any small creature, but to an entomologist it means something much more specific. Bugs, or 'true bugs' as they are sometimes called, in an attempt to avoid confusion, properly belong to one group of insects, the Hemiptera, a collection of creatures that for the most part do not trouble humans much or attract our attention. A few of them are pests – for example, greenfly and mealy bugs – but most live out their lives quietly in meadows and woodlands, feeding on plants or other insects. There are more than fifty species in the meadow at Chez Nauche, including froghoppers, shield bugs (known as stink bugs in the US, for their unpleasant defensive secretions), assassin bugs, whitefly, scale insects and many others. In the pond there are still more: backswimmers, water boatmen, pond skaters and water scorpions (not scorpions at all, but so named because they have a long tail – in actual fact the snorkel through which they breathe).

Although they are a diverse group, all true bugs have one thing in common: their mouths are shaped into a sharp, flexible tube that can be stabbed into their chosen food and used to suck out the juices. Some, such as aphids, use this to suck on plant sap, while others such as the pond skaters and assassin bugs suck the juices from other insects, draining them dry and discarding the empty husk.

Firebugs are not unusual among the true bugs in their enthusiasm for copulation. Many bugs stay bound together for days, with the male and female facing in opposite directions, but joined by the tips of their abdomens. Movement is difficult, and when they are frightened by my weeding and strimming they try to run in opposite directions, with the bigger, stronger female usually getting her way. They remain locked together for the same reason as the dragonflies: the males are trying to ensure their paternity by preventing their partner from mating again. The females may have little choice in the matter, but perhaps have no strong incentive to escape, for if they do they will doubtless then face endless harassment from further suitors.

Bizarrely, firebugs have been discovered to be affected in a most peculiar manner by exposure to American newspapers. Back in the 1960s American researchers imported young firebugs from Europe for their experiments. Instead of growing into adult insects, these firebugs developed into super-sized youngsters that retained juvenile characteristics and were unable to reproduce. This never happened when the bugs were reared in Europe. The scientists eventually deduced that the cause of the problem was the paper they were using to line the rearing boxes. Bugs reared in boxes lined with American newspapers, such as *The New York Times* or *The Washington Post*, remained stuck as juveniles, while if they were reared on imported copies of *The Times* from the UK they were fine. It eventually transpired that a number of American fir-tree species used in paper manufacture contained a chemical that mimicked a hormone in firebugs and so caused their abnormal development.

Although most true bugs lead innocuous and inoffensive lives, this group includes perhaps my all-time least-favourite insect. Many years ago, when I was a poverty-stricken PhD student, I clubbed together with an old university friend and we bought a house. At the time, in 1988, house prices were rocketing, and we were worried that if we didn't get on the housing ladder soon, we'd never be able to (had we but known it, the boom was shortly to be followed by a dramatic crash, but we were young, foolish and impatient). Even with our pooled resources we couldn't afford anything in Oxford, where I was based, so instead we bought an ugly concrete ex-council house in Didcot, twenty-five kilometres to the south. Unless you are a train-spotter – for it is a major railway junction with a railway museum – Didcot is a rather dismal place, but with understandably cheap housing. We furnished the place with second-hand furniture from charity shops and rented out a room to help cover the bills.

After a few months our lodger, a cheerful, pot-smoking, long-haired biker named Mark, started complaining of a rash. He had itchy bumps in meandering lines along his arms and torso. It was summer and we initially blamed mosquitoes from the pond that I had dug in the garden, but as autumn set in and the mosquitoes should have been in decline, the rash became worse. Mark went to see a doctor, but he was unable to diagnose the cause. Mark became convinced that there was something in his room that was biting him – perhaps fleas – but we searched his mattress and could find nothing; and we had no dog or cat at the time. It was all rather unpleasant, and I started itching in sympathy. The rash got worse, and Mark took to leaping out of bed at night and searching his bedroom for the culprit (I would be woken by him crashing about, cursing), but to no avail. Eventually the poor guy moved out, and it was only then that I discovered the cause. I decided to redecorate his room, and in doing so I took down an old wooden shelf. When I unscrewed it from the wall I found dozens of empty, shed skins of some sort of insect, lodged in the narrow crack between the woodwork and the wall. They looked vaguely familiar from old photographs that I had seen as an undergraduate – photos of odd experiments on bugs performed in the 1930s by an entomologist named Vincent Wigglesworth. I started a more thorough search of the room. The bed I had bought had a sturdy old oak frame, and around the joins in the woodwork I found tiny specks of what appeared to be dried blood. There were a few narrow cracks and, peering into them, I thought I could detect movement.

By this point my suspicions were seriously aroused, and so I set about smashing the bed to pieces. When I cracked open the wooden joints, dozens of flat amber insects were revealed and began groggily scurrying about in search of somewhere to hide from the daylight. I realised they were some sort of true bug, and a few moments searching through my books identified them as bedbugs, specialist human blood-suckers. They are nocturnal, sneaking out in the small hours to suck human blood. As the blood at each bite-mark starts to coagulate, they move on a little, giving rise to the distinctive lines of bites that Mark had been suffering from. There were hundreds of them, but their ability to flatten their bodies and squeeze into the tiniest cracks had prevented us from seeing them. Somehow the bedbugs had evaded Mark's night-time searches, too. They were everywhere: in the cracks in the cupboards and even behind the skirting boards. I took all the furniture from the room outside and burned it and then had the whole house dosed in insecticide, one of the few times in my life when I have resorted to using it. Even then some bedbugs survived, and I had to have the house sprayed a second time, at twice the dose, to finish them off. It turns out that bedbugs have made quite a resurgence in the UK in recent years with the advent of central heating, which keeps them cosy and active right through the winter. They have also evolved a degree of resistance to insecticides, which explains why they weren't killed off by the first spraying.

I later looked through my undergrad notes to remind myself of the lectures I had received on Wigglesworth's experiments. Sir Vincent Brian Wigglesworth, to give him his full title, was a great pioneer of studies of insect physiology and the hormonal control of moulting and growth; he is one of very few entomologists ever to receive a knighthood. Wigglesworth served in the army during the First World War, before studying natural sciences at Cambridge, where he eventually became a lecturer and professor. Quite early in his career he discovered several of the major hormones found in insects, including the juvenile hormone whose analogues are found in paper made from North American trees. His discoveries were underpinned by macabre experiments that he conducted on bedbugs and their relatives.

To demonstrate the presence of a hormone in the body fluids of insects that controls moulting, he decapitated bugs and joined their headless corpses with short lengths of glass 'microcapillary' tube – the same type of fine tube that I commonly use to extract nectar from flowers. Once joined in this way, the pair of bugs are strongly reminiscent of a mating couple, except that instead of being tail to tail, they are joined neck to neck. The tubing connects their body fluids, so that any compounds within can flow or diffuse from one body to the other. Although the bugs might reasonably be described as dead at this point, for they have no head or central nervous system (brain), nonetheless they continue to function for many days. As you might imagine, they don't do a great deal – without a head, neither body will decide to go for a walk, and they obviously can't feed. However, their tissues are still alive, they continue to respire and their bodies can moult, shedding their skin and thereby producing the cast skins that I had discovered in Mark's room. Wigglesworth found not only that a decapitated bug could moult, but that if he decapitated a bug that was close to moulting and joined it up with a slightly smaller bug that was nowhere near moulting, then both would moult together. Something was being transferred from the larger to the smaller body, causing it to moult prematurely.

Wigglesworth focused particularly on bedbugs and their relatives, the kissing bugs, and tried joining the different species together. He found that the hormone that controlled moulting was common to both, so that conjoined corpses of the different species also moulted in synchrony. Similar hormones are now known to be found in all insects, although it was many more years before their chemical structures were elucidated.

One of the few positive things to be said for bedbugs is that they do not spread disease. Many biting insects do; most famously, mosquitoes spread malaria and a host of other life-threatening diseases, such as dengue fever and yellow fever. Aphids offer a similar service for plant viruses – between the many types of aphid they spread at least 150 different viruses, many of them very harmful to crops, including such colourfully named diseases as beet mosaic, cherry ringspot, onion yellow dwarf and tomato-spotted wilt. One of the relatively few examples of a true bug that spreads human disease is the kissing bug of South America, the insect that Wigglesworth had imported to conduct his experiments. These large brown insects, about two and a half centimetres long, live in crevices is rural houses and huts. Like bedbugs, they sneak out at night to feed on sleeping humans, where they prefer to feed on the soft, delicate skin of the lips, which is easiest to penetrate. It is from this sinister habit that they get their name. While feeding they often defecate, and inevitably some of their faeces fall into the mouth of their unfortunate sleeping host. Even when the kissing bugs do not feed on the lips, they can infect their unfortunate host, for their bite-marks cause itchy lumps, and the scratching they elicit can lead to the insect's faeces penetrating the wound. The faeces contain virulent spores of *Trypanosoma cruzi*, a single-celled protozoan related to the parasite that causes sleeping sickness in Africa. The South American parasite causes Chagas' disease, an unpleasant chronic illness that is thought currently to infect about eleven million people in Central and South America. Many people suffer no symptoms, but about one-third develop inflammation of the heart and sometimes also the gut, with the heart damage eventually causing death. It seems that Charles Darwin was the unlucky recipient of a kiss from one of these bugs when visiting South America on a stop-off from *The Beagle*'s voyage. In his diaries he describes being bitten while exploring near Mendoza in Argentina:

> At night I experienced an attack, & it deserves no less a name, of the Benchuca, the great black bug of the Pampas. It is most disgusting to feel soft wingless insects, about an inch long, crawling over ones body; before sucking they are quite thin, but afterwards round & bloated with blood, & in this state they are easily squashed.

For much of his later life he suffered from a range of symptoms that were never diagnosed, but which approximate to those caused by Chagas' disease – the disease was discovered by Carlos Chagas some thirty years after Darwin's death. Some have suggested that Darwin was simply a chronic hypochondriac, the condition perhaps brought on by worry at the prospect of publishing his theory of evolution by natural selection, but it may be that the poor chap was genuinely suffering from a life-threatening illness.* There have been moves to test Darwin's remains for DNA fragments of the parasite, but the authorities of Westminster Abbey, where he lies entombed, have so far refused permission.

If bedbugs and kissing bugs sound as if they are best given a wide berth, their cousins the African bat bugs are arguably more gruesome still. These creatures closely resemble bedbugs, but inhabit caves in East Africa, where they feast upon the blood of bats, being active in the daytime when the bats are asleep. In a

---

* The many eccentric cures that Darwin tried included wrapping himself in bands of copper and having his butler pour buckets of ice-cold water over his head.

wonderful irony, they have South American cousins that suck the blood of vampire bats. The mating habits of the African bat bug are perhaps amongst the most barbaric yet discovered in the animal kingdom, and should perhaps cause human females to reflect that, no matter how clumsy their lover, matters could be much worse. The penis of the African bat bug is not dissimilar to its mouthparts – a sharp, pointed tube. Instead of inserting this into the genital opening of the female in the conventional manner, the male bat bug simply grabs the female and stabs his penis through her body wall, injecting his sperm directly into her body cavity, from where it swims to fertilise her eggs. Females are forced to mate many times in their lives, and so can accumulate considerable tissue damage from the multiple stabbings. To make matters worse the male's penis is far from clean – personal hygiene not being a high priority in the bat bug – and so it introduces bacteria into her body, which can lead to infection and death.

In an attempt to combat this, female bat bugs have evolved a fake 'genital opening', a funnel on their back that tries to guide the sharp penis of the male into a cluster of immune cells, which mop up the bacteria. Females still sometimes get stabbed elsewhere, but the damage is reduced.

This sordid story has a further twist. As in many animals, the males are not terribly discriminating in their courtship. They frequently grab and stab other males, causing them considerable damage in the process. As a result, some males have also evolved a fake genital opening to try and minimise the damage, but this in turn makes them look a little more like females and so increases the frequency with which other males try to mate with them. It seems there is no escape for either male or female bat bugs from the damaging sexual depredations of the males of the species.

In marked contrast to the degenerate bat bugs, some true bugs eschew sex altogether for much of the year. Some of the flowers in the meadow – notably including white campions, creeping thistles and meadow vetchling – are commonly attacked by blackfly, sap-sucking aphids. In chapter twelve we will look at why most animals have both males and females and reproduce sexually. Aphids are one of the interesting exceptions, at least during the summer months. The clusters of blackfly on thistles are predominantly wingless females. They plug themselves into the phloem, the network of tubes that transports sugar-rich sap around the plant, and then rarely move again unless they are attacked or the plant dies. They squirt out youngsters at a prodigious pace (up to twelve per day) and these are genetically identical copies of themselves – a process known as parthenogenesis. If sheep could do this, it would have saved scientists the enormous effort and expense that went into producing Dolly (the first artificially cloned mammal). The offspring walk a few millimetres from their mum, plug themselves in and repeat the process. In some aphid species, offspring have their own young already developing inside them as they are born, recalling Russian nesting dolls that are packed one inside the other. Hence one aphid quickly becomes thousands, all identical females descended from a single individual. They sacrifice the advantages associated with sexual reproduction – the mixing of genes – for extremely rapid reproduction.

Only when summer ends and they find themselves under heavy attack from predators, or the plant starts to weaken under the load of aphids, do they change strategy. When this happens, they start to produce both males and females; and what is more, these offspring have wings. They fly away, mate (and so jumble up their genes once more) and start fresh colonies. In autumn they tend to set up their new colonies on woody plants in the hedgerows, where they can survive the winter in relative safety.

The odd clonal nature of aphid colonies has led to the evolution of some remarkable behaviour. In 1977 the Japanese entomologist

Shigeyuki Aoki discovered the existence of soldier aphids, a specialist caste of aphids that exist in some species and defend the colony against predators such as ladybirds. Since Aoki's initial discovery, soldier castes have been discovered in forty or so species of aphid. Oddly, they seem to be particularly common in gall-forming aphids, species that stimulate their host plant to produce a protective, hollow ball of plant tissue, within which they live and feed in a central cavity (one might imagine that aphids living within a gall have less need of soldiers than those living in the open). These soldiers are larger than their genetically identical sisters, and have exaggerated, powerful forelegs and sharp horns on their heads. The soldiers generally don't reproduce themselves, instead selflessly devoting their life to defending their sisters. They sit at the edge of the colony, and if a predator such as a lacewing attacks, they rush in to defend their siblings, attempting to grab the predator and impale it on their horns or stab it with their sharp mouthparts.

Even more remarkably, another Japanese scientist named Takema Fukatsu recently discovered that soldier aphids will also act as paramedics to their host plant. If a caterpillar chews a hole in the gall in which the aphid colonies live, a team of soldier aphids gather round the breach and eject their own gooey body fluids into the gap, mixing and kneading them with their legs until they dry and harden into a scab. The aphid's juices seem to contain an unknown substance that stimulates the plant tissues to grow back neatly over the scar, something that doesn't happen if no aphids are present. Usually many of the soldiers get stuck in their scab and die, their corpses entombed by the growing plant tissues, but Fukatsu found that their sacrifice was effective: aphid colonies in unrepaired galls were rapidly overrun with predators and wiped out, whereas the vast majority of colonies in repaired galls survived.

Such altruistic behaviour is extremely rare in nature. It has an obvious parallel in ants and bees, where workers are sterile and will often sacrifice their lives in defence of the nest. The reason that these two groups show such behaviour lies in the peculiar patterns of genetic relatedness that both show. In most species of animal (including humans) siblings share 50 per cent of their genes. In evolutionary terms, this means that we should care about our own survival and success twice as much as we care about that of our sisters and brothers. Given the choice between saving our own skin or saving a sibling, we should save ourselves every time. An informative, if rather silly scenario is to imagine what you would do if you were kidnapped, along with an assortment of your relatives, by terrorists. Suppose the terrorists offer you a choice: they will shoot you, or your sibling. Genetically speaking, you should sacrifice your sibling; after all, he or she only carries half of your genes. If the deal offered is a choice between your own life and that of two of your siblings, then in evolutionary terms it makes no difference which choice you make. But if you could save three siblings, then *you* should take the bullet; together your siblings have 50 per cent more of your genes than you do. Similarly, you should cheerfully sacrifice seven cousins rather than die yourself (cousins each having one-eighth of your genes), but you should willingly give your life to save nine of them.

Of course I'm not suggesting that humans, aphids or bees actually think about it in these terms; but we would expect natural selection to favour individuals with behaviours that approximate to these predictions. If you think that humans, with our capacity for thought and reasoning, have risen above such primitive urges, ask yourself this: who would you be most willing to risk your life for, a close relative or a distant one? Who will you leave money to when you die? Many wills divide up assets so that the bulk goes to the closest relatives, and smaller sums to more distant ones; they reflect patterns of genetic relatedness.

I am perhaps getting a little off the point; aphids and bees don't make wills, and they do not generally get kidnapped. What they do have is unusually close relatedness between members of groups. In female ants and bees, sisters share 75 per cent of their genes, which makes cooperation and self-sacrifice between them more likely to be worthwhile. Female aphids within a colony share 100 per cent of their genes – they are all identical, which makes it easy to understand why they might risk their lives to defend their sisters. If a soldier aphid can save the life of just one sibling by sacrificing itself, it has broken even. If it can save the whole colony, then in evolutionary terms it has made a very wise move indeed.

Most aphids don't have a soldier caste, but instead enlist the help of other animals in defence of their colony. The blackfly colonies on the vetchling and thistles at Chez Nauche are assiduously guarded by ants. The ant workers patrol among the aphids almost as a farmer might watch over a herd of cows; if a hungry ladybird attempts to snaffle an aphid or two for lunch, the ants angrily attack, simultaneously emitting an alarm pheromone that calls more of their sisters to their aid. They will chase and bite the ladybird until it gives up and flies off to try its luck elsewhere. The cow analogy is particularly apt, since the ants are not caring for the aphids out of the kindness of their heart; they milk them for sugary honeydew. This is a good deal for the aphids, as they have more sugar than they need. Plant sap is rich in sugar, but generally quite low in protein, so aphids have to drink a lot to obtain enough protein. This gives them a surplus of sugar, which in the absence of ants they simply excrete; if you have ever parked your car under an aphid-infested tree you will be all too familiar with the sticky spots of honeydew that rain down from the aphids above. Ant–aphid mutualisms of this sort are found all over the world, and some have become more complex. In some species the ants will pick up and move the aphids to another plant if the one they are on starts to die. Other ants take the eggs that many aphids lay in autumn into their nest for the winter, keeping them warm and safe (like cows in a barn) until the spring, when they carry the newly hatched nymphs back out and place them on a suitable plant. At Chez Nauche, yellow meadow ants keep several species of aphids permanently underground in their nests, where the aphids feed on the roots of plants and are milked by the ants.

Of course the care shown by the ants to their aphid charges is entirely selfish. When aphids switch to their winged phase, ants have been seen to rip their wings off to prevent them flying away, just as a farmer might clip the wings of his poultry. When food is scarce in winter the yellow meadow ants consume their aphids, but generally keep enough safe that they can build up their stocks again in spring.

If aphids use sugary sap to buy the aid of ant guards, other true bugs use their plentiful supply of sap to defend themselves in quite a different way. In spring one of the hazards of walking though the long grass in my meadow is that your legs are likely to become liberally splattered with cuckoo spit. Cuckoo spit is not, of course, the spit of a cuckoo, and the origins of this peculiar name are lost in time, although it does appear at about the same time that migratory cuckoos arrive from Africa. In fact it is the frothy secretion of young froghoppers – rather cute, chubby little bugs, which as adults live up to their name by performing astonishing leaps when disturbed.* As youngsters they are much more sedentary, and spend their time hidden within the centre of their own personal bubble bath, which they excrete from their rear end and whip up

---

* The leaps of froghoppers are amongst the most impressive of all insects'. They can accelerate themselves from a standstill at 4,000 metres per second, subjecting themselves to 400 'Gs' – the force of gravity – in the process.

as necessary with their hindlegs. The glutinous mass provides excellent protection from almost all predators, including birds, which could easily find the conspicuous white blobs in the meadow, but seem to be unwilling to get their beaks covered in the slimy, bitter-tasting froth in order to pick out the tasty nymphs inside.

True bugs are just one relatively small group of generally inconspicuous insects. Keen gardeners will notice aphids from time to time, and if you are unlucky enough to get a bedbug infestation it will be hard to overlook their presence for long, but otherwise it would be easy to live your whole life in ignorance of these creatures. Some may have seemingly grotesque and unpleasant habits, while others will selflessly give their lives to save their sisters. Their headless corpses have played a role in helping us to understand how insect hormones work; and one hungry bug may even have killed Darwin. Yet most true bugs are unstudied and unknown, apart perhaps from a pinned specimen in a museum. We have barely scratched the surface of this topic. What even more fascinating natural history has yet to be discovered among these remarkable insects?

PART II

The Rich Tapestry of Life

The lives of the myriad creatures of the meadow are, as we have seen, fascinatingly varied, sometimes unexpected and often unknown. However, the greater mystery is how they all fit together. No creature lives in isolation – all are interlinked one way or another in a dynamic web of interactions that we are only just beginning to explore. The bedrock of meadow life is the plants, which capture energy from the sun and transform it into food that ultimately supports more or less everything else. The plants themselves compete for space, light, water, nutrients and pollinators, have mutualistic relationships with bacteria and fungi, and are attacked by diseases, parasites and herbivores. What determines how many species there are? What happens if species are lost, or new ones invade? Why are some animals and plants rare, while others are common? How many species are necessary to maintain healthy soils, and to ensure adequate pollination? More often than not we do not have answers to these questions.

Here I will explore some of the ways in which the creatures and flowers of the meadow interact with one another, focusing particularly on pollination, a process that underlies much of the diversity of both plant and insect life on Earth. Pollination might seem to be a simple, harmonious activity – bees buzzing from flower to flower, drinking nectar and carrying pollen – but look closely and a web of deception, competition and robbery is revealed. It is these interactions between species that form the rich tapestry of life.

CHAPTER TEN

Hothouse Flowers

15 September 2011. Run: 38 mins 50 secs. People: 4 – hunters with shotguns under their arms and slaughter on their minds. Dogs: 5 – I armed myself with a stick before passing the spaniel's house, which seemed to deter attack. Butterfly species: 9, all rather tired, as summer comes to an end. I glimpsed a hoopoe in the hamlet of Le Breuil, using its long, downcurved beak to probe for insects in a cowpat on the road through. They are common enough, but I can never fail to get excited by this splendid clown of a bird with bright-orange, black and white markings and a foolish crest of feathers. It headed off in characteristic swooping, woodpecker-like flight when it saw me.

There is endless and wonderful variety in the structure of flowers, and I would encourage anyone who has never taken the time to look at them carefully to do so. Although we spend much energy in growing pretty flowers in our gardens, and expend considerable sums on buying cut flowers for our loved ones, the pleasing forms of flowers are not for our benefit. Their complex shapes and fabulous colours are the culmination of 130 million years or so of co-evolution between pollinators and plants. Each flower is both an advert and a trading platform. The purpose of flowers is to attract pollinators – generally insects – and then to persuade them to carry pollen grains to another flower of the same species, but on a different plant, in exchange for rewards. This is no simple matter, for there are many other flowers of the same and different species, all competing for the attention of pollinators. There are also many different types of pollinator. At Chez Nauche there are a dozen or more species of bumblebee, honeybees, probably in excess of fifty species of solitary bee (I am still trying to identify many of them), numerous butterflies, moths, beetles, hoverflies, and so on. In more exotic climes pollinators may also include bats, lizards, various birds such as parrots and hummingbirds, even mammals.* Each pollinator has a different shape, size and tongue length, and some are only active at certain times of the day or night, or at particular times of the year. For example, some solitary bees are only on the wing for a few weeks; the hairy-footed flower bee, *Anthophora plumipes*, appears in March or April and is gone by May.

From a pollinator's perspective, there may be a plethora of flowers in a meadow. How does it choose which ones to visit, and in what order? Does it go for the commonest, tallest, largest, prettiest or most fragrant ones? It also has somehow to take into account what all the other pollinators might be doing, for if they all make the same choice, then they will all end up fighting over the same flowers, which would not be productive for any of them. For bees it is particularly important to maximise the amount of food they gather, for this directly determines the number of offspring they can rear. From the plant's perspective,

---

* Probably the best-known mammalian pollinator is the incredibly cute honey possum of south-western Australia. These tiny, long-snouted creatures are the only mammals to feed exclusively on nectar and pollen, having a brush-like tip to their tongue to aid in pollen collection. Amongst mammals they are unique in a number of other ways: they have the largest testes, proportional to their size, and the smallest young at birth, weighing just 1/200th of a gram.

which insects should it try to attract? Each varies in its abundance, size, speed and effectiveness as a pollinator. Should it expend lots of resources on big, brightly coloured petals to advertise its presence, but then have few resources left to provide rewards for its pollinators, or will insects see through such unfounded marketing hype? It may be better to pour resources into providing rich rewards and spend little on advertising, relying on discerning, intelligent insects to discover its worth. There is no one answer, and the meadow at Chez Nauche is packed with different flowers, each of which has adopted a slightly different approach. A meadow filled with flowers and busy insects is a complex web of interactions, which shifts through the seasons as different flowers come into bloom and different suites of insect pollinators come and go.

Some insects choose a generalist strategy, flitting readily between different types of flowers. Others specialise, and have often evolved particular structures to help them deal with particular types of plants; in particular, some bees have evolved long tongues to help them gather the nectar in deep flowers that many other insects cannot reach. This approach limits their flower choices, for long tongues are unwieldy for feeding on shallow flowers, but it also cuts down the competition to only those other insects with long tongues. Correspondingly, some flowers are generalists, aiming to attract any old pollinator they can – the hogweed growing along hedge banks and the wild carrot that flourishes in the open meadow are good examples. They do not hide their rewards, but display them within easy reach on a broad, flat platform of tiny white flowers, each a shallow dish of nectar with pollen on offer on short stamens above. The rewards can easily be gathered without any special apparatus; such flowers tend to attract a broad variety of flies, beetles and short-tongued bees. The advantage of this approach is that the flowers are likely to get lots of insect visits, but there is a downside: their visitors will themselves tend to be generalists, and so they are likely to flit off to a different flower species for their next drink of nectar. Pollen transfer between different plant species is of no use to either the donor or the recipient. In fact the female part, the stigma, can become clogged up with pollen if too much arrives from a different flower species, leaving no room for the correct pollen, should it subsequently be delivered.

Much of the strategy adopted by a particular flower species can be discerned by close examination of its structure. Almost all flowers are constructed of essentially the same parts, though they may differ wildly in shape, size and colour between species. The most obvious parts are usually the brightly coloured petals, designed to attract pollinators from far and wide. Flowers that are aiming to attract bees are often yellow or purple, since these are the colours to which a bee's eyes are most sensitive. The petals of bee-pollinated flowers often have nectar guides in ultraviolet, invisible to us without special equipment, but obvious to bees, which lead them to the rewards. In some bee-pollinated flowers, such as clovers, foxgloves and dead-nettles, the petals form a deep tube with the nectar at the bottom, so that only long-tongued bees can reach it. Often these flowers have bilateral symmetry (rather than the radial symmetry found in the majority of flowers), a design that seems to be inherently appealing to bees.

Many flowers target bees because they are the most assiduous and hard-working of pollinators, driven by the need to feed their young, but some plants aim to attract other insects. Butterfly-pollinated flowers tend to be pink or red, although there are few of these in Europe. Moth-pollinated flowers, such as white campions, tend to be white so that they can be seen more easily in the dark, and they typically have a strong scent; honeysuckle is another example, which clambers up the small elms and blackthorns along the drive at Chez Nauche, and is a favourite for growing up garden trellises because of its evening fragrance. Fly-pollinated flowers are often bowl- or plate-like – think of the yellow dish of buttercups, or the white platform provided by meadowsweet and elder flowers. In these, the nectar is readily available to sip, even for insects with short tongues. Some fascinating fly-pollinated plants stink of rotting meat to attract carrion flies, though fortunately there are no such plants in my meadow.*

Although the petals are often the most obvious part of a flower, the design of the rest of the structure is equally important to its success. Inside the petals – or sometimes protruding from them, where they will contact approaching pollinators – are the male and female parts. Often the sexual parts are hidden so that only pollinators of the correct shape and size can find them. Meadow clary, a wild relative of garden salvias, has anthers hidden amongst the upper petals in its purple flower, and a nifty mechanism whereby the action of a bee probing the nectaries activates a lever, causing the anthers to swing down and place pollen precisely on the thorax of the insect, just behind its head. There was no meadow clary in my meadow until recently, for it has heavy seeds and so takes a very long time to recolonise suitable areas, but I collected some seeds locally and sprinkled them into the grass, and I now have a handful of these lovely

---

* The largest flower on Earth is that of the rare plant *Rafflesia arnoldii*, which is found in the dense, steamy rainforests of Borneo and Sumatra. The brownish-pink, mottled flowers are about one metre across, can weigh more than ten kilograms and reek of decaying flesh to attract flies. This plant is also peculiar in having no leaves and little in the way of stem or roots; it is a parasite, sucking its nutrients from rain-forest vines.

plants coming through. I can never resist plucking a slender grass stem and gently inserting it into the mouth of the flower, mimicking the action of a bee and causing the anthers to swing gracefully down.

Broom has anthers hidden in the bottom of the flower, which swing upwards when a bee of sufficient weight lands, covering her tummy in pollen. Because of this mechanism, broom is largely pollinated by bumblebees, and does not waste its pollen on other less reliable, lightweight pollinators. The trigger plants of Western Australia have spring-loaded anthers, which, when triggered by the arrival of a bee, swing round at high speed like the arm of a mousetrap to smack pollen on to a particular part of the bee. Each species of trigger plants uses a different part of the insect, so their pollens do not get mixed up; some place pollen on the bee's back, some on her belly, some on her left side, some on her right.

Some bee-pollinated flowers, such as woody nightshade, hide their pollen inside tubular anthers with just a tiny hole at the bottom, from which the pollen can only be extracted by violent shaking, something known as 'buzz pollination'. Only a few bee species have evolved the ability to do this. Honeybees, for all the sophistication of their colonies and their communication mechanisms, seem not to have mastered the technique. In contrast, bumblebees and some solitary bees are adepts; they grasp the anthers with their jaws and then use their wing muscles to vibrate their body, shaking the whole flower and causing a shower of pollen to fall from the hole in the anthers, which they expertly catch with their hairy legs. Other plants, such as the creeping thistle that spreads in the more fertile parts of my meadow, take the opposite tack; they dispense with subtlety entirely and simply produce heaps of pollen, almost drowning visiting insects in a sea of anthers, so that they become white with sticky pollen. Similarly, large *Banksia* flowers and many other bird-pollinated flowers in Australia have numerous anthers on long stalks protruding from the flower, so that any bird probing with its bill for nectar has pollen dusted all over its face.

No matter how beautiful the flower, it is not likely to attract insects unless it provides some kind of reward. Many flowers use nectar as their reward; this is simply sugary water, sometimes laced with a few traces of amino acids and other nutrients. Adult insects need fuel to fly, so that they can find mates or places to lay eggs, and nectar is the perfect high-energy drink to keep them in the air – Lucozade for bees. Because plants can create sugars from just water and carbon dioxide by photosynthesis, producing nectar is not enormously costly for them, and some plants produce heaps. Borage and comfrey are both prolifically rich in sweet nectar, and so they are very popular with bees. This presumably gives these plants a distinct advantage in the competition for pollination services.

Other plants produce no nectar, but instead use pollen to entice insects. Pollen is of no interest to some insects, such as most butterflies* and moths, because they can only cope with liquid diets; they are short-lived creatures, designed to quickly find a

---

* In the 1970s Lawrence Gilbert, a butterfly expert from the University of Texas in Austin, discovered that the *Heliconius* butterflies of South America are able to digest pollen. These elegant, long-winged neotropical insects are Methuselahs of the butterfly world, living for up to six months as active adults, whereas most butterflies live for just a couple of weeks. To fuel this longevity they collect a ball of pollen at the base of their tongue and exude sugary liquid on to it. This is sufficient to cause the pollen to release much of its amino acids (the building blocks of proteins), which the insects can then suck up through their tubular tongues.

mate and reproduce. However, pollen is consumed by some longer-lived insects, including various beetles and hoverflies, and of course by bees, for whom it is a vital protein source for adults and especially for their larvae back in the nest. The protein content of pollens varies greatly, with some plants such as legumes (clovers, vetches, trefoils, peas and beans) producing particularly protein-rich pollen that helps them to attract bees. That is part of the reason why I have been so keen to encourage these plants at Chez Nauche.

It has relatively recently been discovered that the rewards provided by flowers can go beyond the nutritive. Some flowers, including magnolias and some species of lily, generate their own heat. In springtime cuckoo pints, also known as lords and ladies, are common in the hedge bottoms at Chez Nauche. These are a type of arum lily, with peculiar and striking flowers consisting of a purple, poker-shaped vertical rod, called a spadix, partially enclosed in a green hood, the spathe. The male and female parts are clustered at the base of the spadix, beneath a ring of hairs. The spadix produces heat, warming to 15°C above the ambient temperature; they are quite hot to the touch. This helps to evaporate the floral odours, which are distinctly unpleasant and vaguely dungy. Unappealing though they may be to us, they attract pollinators, mainly tiny flies known as owl midges or moth flies. These flies normally breed in decaying organic matter, including sewage; they can even breed in drainpipes and U-bends, and often turn up in the bathroom at Chez Nauche, providing food for the many spiders. These flies are tricked by the plant, for it provides no breeding ground, and they become trapped within the lower part of the spathe, beneath the hairs on the spadix. Once the female parts are pollinated, the plant releases pollen on to the flies, and the hairs wither, allowing the flies to escape. More likely than not they will be tricked again by another cuckoo pint, for they are not the smartest of insects, and so pollen is transferred from flower to flower. In this instance the heat production is purely for the plant's benefit, and the insects gain no reward whatsoever, but in some plants the heat is used to benefit the insect. For example, flowers of the tropical arum lily *Philodendron* can be an astonishing 30°C warmer than the ambient temperature. They attract scarab beetles, which settle, feed and mate upon the lily. The flower folds around the beetles, trapping them in place for a few hours, although in this instance the beetles are presumably happy enough, for they are warm, have plenty of food and spend most of their time copulating and eating at the same time. Once the flower is thoroughly pollinated by all this beetle activity, it releases its own pollen, along with a burst of heat to warm the beetles, before unfolding to allow them to depart, covered in pollen.

Others flowers have evolved structures and mechanisms to help them trap heat from the sun, acting as miniature hothouses. Many bowl-shaped flowers become much warmer than the surrounding air. This effect can be greatly enhanced by heliotropism, whereby the flower slowly rotates through the day so that it is always facing the sun. Probably the best-known example is the sunflower, commonly grown as an arable crop in the Charente. The plants are sown in serried rows across fields of many hectares and, when in bloom in July every single plate-like flower is aligned towards to sun. This makes them particularly photogenic, but it serves a dual purpose for the plant: both warming their pollinators and aiding the germination of pollen grains and fertilisation of the ovules. Heliotropism is particularly common among Arctic and high-alpine plants, where insects may be scarce and temperatures low; for example, flowers of the Arctic poppy keep themselves 6°C warmer that the surrounding air by tracking the sun in this way.

Some fascinating studies carried out by Beverley Glover's research group at the University of Cambridge have shown that flowers have even evolved special conical cells on the surface of their petals, which focus light like thousands of tiny magnifying glasses on to the cell vacuoles, which are packed with pigment to absorb the heat. These cells are found in Arctic poppies, in spring-flowering crocuses and in snapdragons, which have been a particular focus of their work. The snapdragons have a great advantage as a research tool because a mutant version exists, known as *mixta*, which lacks the conical cells. This makes it easy to compare the temperature of flowers that are identical in every other way, and to study how bees respond to them. Flowers with conical cells are indeed warmer – up to 8°C warmer than the *mixta* mutants – and they are preferred by bees. In field experiments with flowers of both types from which the anthers had been removed (to stop self-pollination), the *mixta* mutants set far fewer seeds because they received fewer visits from bumblebees. However, it has recently become clear that the reason bees prefer flowers with conical cells may have more to do with the extra grip they provide than with their temperature; when landing on the *mixta* mutants, bees have trouble keeping their footing and slip about, sometimes falling off the flower entirely, whereas the texture provided by the conical cells gives them a firm grip. It is not yet known how common such specialised, grippy petal cells are in wild flowers.

Why might bumblebees prefer warm flowers? Bees need to keep warm to stay in the air. Bumblebees are particularly fat insects with small wings, so they have to beat them 200 times per second to stay aloft. This can only be done if their flight muscles are warm – at least 30°C – which is hard to maintain, particularly in early spring when the ambient temperature may be just above freezing. They have to shiver to get warm enough to take off; once in the air, flying itself creates a lot of heat, and of course their furry coats help to keep the heat in, but nonetheless as soon as they land on a flower to feed they start to cool, and if they sit still for too long they have to start shivering again to warm up, all of which wastes a lot of energy. Hence flowers that are warm are particularly attractive to bees, for this helps to keep them warm while feeding. What is more, a warm flower will have warm nectar: a hot, sweet drink, just the thing for a busy bee on a cold day.

There is a further, alcoholic twist to this story. It has long been known that nectar in plants contains yeasts that are transmitted from flower to flower by pollinators, and which thrive on the sugars, fermenting them to alcohol. It is generally assumed that this does not benefit the plant, and that it may indeed be harmful if too much sugar is used up by the yeast. However, a recent study by Carlos Hererra and Maria Pozo at the Doñana Biological Station in Spain found that the metabolic activity of yeasts in winter-flowering hellebores could raise the flower's temperature by up to 6°C. Thus the hellebores' bumblebee pollinators were treated to a warm, sweet, alcoholic drink – glühwein for bees.

Some flowers seem to offer only thermal rewards. *Oncocyclus* are beautiful, showy purple irises found in the eastern Mediterranean. They seem to have no nectar, and are pollinated entirely by male solitary longhorn bees (which have endearingly enormous antennae) that sleep in the flowers at night. The deep-purple flowers warm up swiftly in the early morning, kick-starting the bees' day and enabling them, quite literally, to get off to a flying start.

The most perverse strategy adopted by plants in the battle for pollination is by those species that have opted to offer no reward whatsoever. This seems to be particularly common in orchids, a plant family that adopts a range of peculiar and unusual approaches to pollination. I have already mentioned the bee orchids that mimic the shape, texture and smell of the female of the longhorn bees that sleep within *Oncocyclus* irises. The female bees lack the long antennae, which the males use to locate females. In truth the flowers are a pretty hopeless mimic of the female bee in appearance. Presumably the smell is more convincing, or perhaps male longhorn bees – in common with the males of many animal species – are simply none too discerning when it comes to choosing a partner for a brief liaison. Most flowers produce loose pollen, but orchids package it into balls known as pollinia, each of which has a white stalk with a sticky tip. These are positioned so that the stalk sticks to a visiting insect, which then spends the next few days with the pollinia glued to its face or the top of its head as it travels around. Each flower produces just two pollinia, and they are readily visible, so it is easy to tell whether a flower has had its pollen removed. I have not yet seen longhorn bees at Chez Nauche, and neither have I ever seen a bee orchid flower with the pollinia removed. Through much of northern Europe bee orchids seem to lack their pollinator and rely upon self-pollination to set seed, something that does not bode well for their long-term genetic health and survival.

Mimicking a female bee is a fascinating but understandable approach to achieving pollination. Harder to explain are the many orchid species that offer neither reward nor the false promise of a receptive female. These species rely on naïve pollinators, usually bees, that have not yet learned to avoid them. This may be aided by flowering in early spring, as many nectarless orchids do, for at this time of year there are few other flowers, and queen bumblebees have emerged from a long sleep into a world where all the flowers they encounter are unfamiliar. Some of the later-flowering nectar-less orchids appear to mimic rewarding flowers that grow nearby.

In alpine meadows in Switzerland I have seen nectarless broad-leaved marsh orchids, *Dactylorhiza majalis*, growing amongst stands of nectar-rich bistort, which they closely resemble. Male bumblebees seem to be particularly easily duped, perhaps because their mind is on other things, and I have seen some males with dozens of pollinia glued all over their face – so many that they were barely able to see.

In Britain nearly half of the common orchid species and 90 per cent of the rare species appear to offer nothing to pollinators. Darwin speculated that the rarity of many species may be precisely because they offer no reward. This notion is supported by recent studies comparing seed set of orchid species that produce nectar with species that don't; the flowers of nectar-producing species set seed approximately twice as often as the nectarless species do. This leaves something of a mystery as to why the nectarless species persist with their miserly strategy.

Orchids are not the only flowers to offer no reward. Even in flowers that generally provide nectar, some individual plants produce none. Indeed, this seems to be the norm; some studies suggest that as many as 80 per cent of plant species that are ordinarily regarded as nectar-producing have some nectarless plants. These plants appear to be cheaters, relying on other members of their species to provide the reward. One can measure the amount of nectar in flowers by using a microcapillary tube, a very thin and delicate glass tube that is inserted gently into the nectary of the flower. If nectar is present, it draws it up via capillary action, and the length of the column of nectar visible within the glass tube is a measure of the volume present. Over the years I have spent many hours trying to get nectar out of different flower species, and it is a fiddly and frustrating business; bees are much better at it than humans. At any one time many flowers will be empty because an insect has drained them, but this can be overcome by 'bagging' them – enclosing the flower for a few hours in a bag of fine netting to exclude insects, so that the flowers can fill with nectar. I once spent some time studying nectar production in centaury in the meadow at Chez Nauche. Centaury blooms in July, producing small, pink star-shaped flowers. I had become interested in the foraging behaviour of the small, solitary bees that seem to be its main pollinator. No matter how hard I tried, from nearly half the plants I could extract no nectar whatsoever, even when I bagged them overnight.

Producing no nectar is an interesting but risky strategy for an individual plant. It has the clear advantage that the plant saves the energy it would otherwise have put into nectar, but it relies on the presence of enough other plants of the same species that are producing nectar to convince pollinators that the plant species is worth bothering with; or else it requires there to be a plentiful supply of naïve pollinators. The success of the cheating strategy is also very much dependent on the intelligence of the pollinators. Bumblebee workers, for example, are pretty sharp operators and are not fooled easily or for long. They spend their life learning and relearning associations between the colours, shapes and scents of flowers, and rewards. When they first begin foraging they experiment with visiting different flowers, and quickly learn to search out selectively those that are most rewarding. They commonly specialise on a particular flower species, be it red clover, tufted vetch or viper's bugloss, and become very adept at extracting the rewards swiftly and efficiently – much as we become better at any task with practice. If you follow a bumblebee in a flowery meadow, you will quickly notice that she usually visits the same flower type over and over again, bypassing most other flowers, a behaviour that was familiar to Charles Darwin in 1876 and has become known as 'floral constancy':

That insects should visit the flowers of the same species for as long as they can is of great significance to the plant, as it favours cross fertilisation of distinct individuals of the same species; but no one will suppose that insects act in this matter for the good of the plant. The cause probably lies in insects being thus enabled to work quicker; they have just learned how to stand in the best position on the flower, and how far and in what direction to insert their proboscides.

As Darwin observed, this behaviour is exactly what the plants want, for it delivers pollen swiftly and efficiently from one flower to the next, without the pollens from different species getting mixed up and deposited on the wrong stigma. The complexity of flowers can be seen as a mechanism to encourage this, a way to manipulate bees into committing themselves to the relationship; if working out how to find and extract the reward from a particular species is difficult, then once the skill is mastered it makes sense to utilise it to the full, not least because competition with other pollinators is likely to be less on such taxing flowers.

All this said, floral constancy only makes sense if the flower that the bee has chosen to specialise on is providing a good reward. Bumblebees constantly reassess their strategy, and occasionally try other flowers that they pass, presumably to check they aren't missing out on something better. I have my favourite cheeses, but every now and then I try something different that I've never tasted before, in case it turns out to be fantastic. Often it is a disappointment – I recently experimented with some Saint Paulin and found it to be tasteless and rubbery, so I'll not try it again. Similarly, a bumblebee visiting viper's bugloss might quickly check out a geranium as it passes. If the geranium turns out to have sweeter or more copious nectar than the bugloss, the bee will try a few more; and if the reward is consistent, she will quickly abandon bugloss altogether. If not, she will tend to stick with the bugloss, but will occasionally experiment with new flowers that she happens to encounter. If the bee encounters a run of empty flowers of her favoured plant, she starts spending more time checking out the alternatives. She does the same if a flower becomes hard to find, such as might happen when the favoured flower is beginning to go over.

Bee foraging behaviour can be seen as a trade-off between the benefit of sticking with one flower type that the bee has become skilled in handling and trying to minimise the risk of missing out on something much better. In this way, bees keep tabs on the options available to them, and will swiftly abandon one flower in favour of another if it provides a better reward.

Bee discrimination among flowers goes beyond discerning which species are most rewarding. Bees are also capable of learning which particular patches of flowers are best. I have spent many happy hours watching the bees, butterflies and moths visiting the lavender bushes along the front of the house at Chez Nauche. After a little while it is possible to start recognising individual bees. For example, you might spot a particularly brightly coloured shrill carder bee. The largest lavender bush, just to the right of the front door, always seems to be the most popular with insects, and that is where this bee is foraging. After ten minutes or so, when her honey stomach is filled, she flies back to her nest somewhere in the meadow, laden with nearly her own body weight of food. Half an hour later she returns, unladen, and this repeats itself through the day. You can almost set your watch by her return. Of course you might suspect that I am imagining it – there could be more than one brightly coloured shrill carder of similar size – but I know that I am not, because I have tried marking the bees with a little blob of correction fluid. This sometimes upsets them and then they zoom off never to return, but usually it doesn't, and the same individual can then be seen coming back at regular intervals, day after day, visiting exactly the same bush.

This in itself wouldn't be too remarkable. The bee has found a good patch of food, remembered its location and is simply flying backwards and forwards from her nest, a trip that takes a predictable length of time. What is perhaps more impressive is that she may be visiting a number of different flower patches on each trip. She can remember the locations of a dozen or more patches and navigate in a straight line from one to another. This behaviour is known as trap-lining, derived from fur-trapping, where the trapper will tour his traps along a fixed route every day. It has been studied extensively by the Canadian bumblebee expert James D. Thomson at the University of Toronto. The bee does not visit the patches in the order that she discovered them, but instead plots novel routes linking the nearest patches together, effectively minimising her flight time. When she can carry no more, she does not need to retrace her steps, but can manage to fly in a direct line back to her nest. It seems that most bumblebee species do this, and also honeybees, hummingbirds and those long-lived *Heliconius* butterflies. Trap-lines can remain stable for many days, or they can shift; if a patch becomes unrewarding, the pollinator skips it and may try out new patches and incorporate them into her route if they are worthwhile.

To return to nectarless flowers, it is clear from all of this that bees are well equipped to swiftly detect and avoid a species of plant that provides no nectar, such as many orchids, or to avoid individual plants (flower patches) that are less rewarding. If nectarless flowers are scarce, they may be visited occasionally by bees that are exploring new patches. Bugloss is generally highly rewarding and is visited by many bees, so you might imagine that a single nectarless plant among many rewarding ones would still receive bee visits. The nectarless plant would effectively be parasitising its neighbours, benefiting from the nectar they produce while contributing none of its own. You might predict that cheating of this sort might spread, for the cheats ought to be at an advantage. As unrewarding plants become more common, bees would be likely to switch away to feed on other plant species entirely, but if so this cost would be borne equally by all of the bugloss plants, and not just the cheats, so the cheating strategy would not be penalised. However, it would seem this does not generally happen – the cheating, nectarless strategy rarely takes over. The explanation may be that nectarless plants, although they may receive some exploratory visits, will never receive the regular, clockwork visits of a trap-lining bee, and hence will not be pollinated as effectively as their rewarding neighbours. In a sense, the intelligence of bees acts to police cheating strategies among the plants they pollinate.

As you will have gathered, getting pollinated is no simple business, with plants competing for the attention and fidelity of pollinators whilst attempting to minimise the costs of doing so. Equally, gathering pollen and nectar efficiently, when it is hidden in varying and unpredictable amounts amongst patchily distributed flowers of numerous colours and shapes, is far from simple. There is doubtless much more to discover about the ecology and behaviour of pollinators and flowers. You could spend a lifetime studying the insects and flowers at Chez Nauche and still have barely begun to scratch the surface. At present we have only a poor idea of how important particular linkages between flowers and pollinators are. What happens if a pollinator species goes extinct? Do others simply take up the slack, exploiting the unoccupied niche, or do some plants suffer from its absence? How many different species of pollinator does it take to support a healthy wild-flower meadow? These questions are pertinent because, as we shall see in chapter thirteen, we are slowly losing our pollinators.

CHAPTER ELEVEN
Robbing Rattle

26 May 2012. Run: 39 mins 48 secs. People: 6 Lycra-clad old men rocketed past me on their bicycles, nearly making me jump out of my skin. Dogs: 5, including an unfortunate three-legged mongrel wandering the street in Épenède. Butterflies: 12 – spotted a male small blue butterfly perched on a grass stem right by my car on the drive, a tiny but aggressively territorial species with smoky-blue wings, the first I have seen at Chez Nauche. This is my favourite time of year; the meadow is in full bloom, the hedge banks lush and bursting with life, without the tiredness that comes in the heat of summer. Several red-shanked carder bumblebee queens were foraging on white deadnettle on the ditch banks; also one shrill carder queen, emitting her distinctively piercing buzz.

> *All, all is theft, all is unceasing and rigorous competition in nature; the desire to make off with the substance of others is the foremost – the most legitimate – passion nature has bred into us and, without doubt, the most agreeable one.*
>
> Marquis de Sade

My first encounter with a truly ancient flower-rich grassland was in Oxfordshire, when I was an undergraduate at Oxford. In some ways it formed the inspiration for my attempts to create a similar meadow in France, for it was in Bernwood Meadows that

I first came to appreciate the extraordinary diversity of life that grasslands can support, if left undisturbed for long enough.

When I went to university I was shy, immature and more than a bit homesick. I was intimidated by the brash confidence of some of the students from public schools, and I struggled to fit in and make friends. My first few terms were a pretty lonely time, and I occupied myself by exploring the local countryside on my old and dangerously decrepit Yamaha 250cc motorbike. There was one place in particular to which I repeatedly returned: Bernwood Forest and the adjacent Bernwood Meadows, about eleven kilometres to the east of Oxford, the site where a few years later I would end up doing much of the work for my PhD on meadow brown butterflies. At the time I didn't really notice the meadow; I walked through it looking for butterflies and other insects, particularly the elusive and rare black hairstreak butterflies that live in the dense blackthorn hedges, but I didn't pay too much attention to the plant life.

It was later, when I had moved 'up the hill' to do my PhD at Oxford Brookes University (or Polytechnic as it was at the time, situated on Headington Hill to the east of the city), that I came to appreciate the meadow for what it is: a tiny surviving fragment of one of the most flower-rich habitats in Britain. The 'Poly' ran an annual field course for biology students in July, rather unexcitingly based in Oxford, but with day trips out to a range of different habitats, one of which was to Bernwood. Just a few months after I had started my PhD I was asked along as a demonstrator on this trip, something that made me very nervous as I wasn't sure that I had any great knowledge to impart. Luckily for me, the member of staff in charge was Dr Andrew Lack, an enormously knowledgeable, twinkly-eyed botanist with a boundless enthusiasm for natural history of all sorts. He walked us through the meadow, pointing out the long, regular ridges and furrows, evidence that the field had not been ploughed since it was cultivated in strips in medieval times. Given 500 years or so in peace, the meadow had had time to develop a wonderful richness of plant life, which in turn supported a huge diversity of insects.

We introduced the students to the many different bumblebee and butterfly species that lived there, and Andrew showed off to the students by catching the male, stingless bumblebees in his hands – a trick that I have since borrowed from him and have used many times to great effect. We spotted lovely bright-yellow crab spiders, perched on the edge of ox-eye daisies waiting for a meal to arrive, and we caught five or six grasshopper species in the long grass. Eventually Andrew set the students to work in small groups, identifying and counting the multitude of different plants within metal quadrats that they scattered on the ground.

I knew some of the plants, particularly those that I had taught myself to identify because they were food plants for the butterflies I had studied and bred during my childhood, but there were many others at Bernwood with which I was not familiar: eyebright, tormentil, lady's bedstraw and pignut, among many others. One of the more common plants in the meadow was one that I had not seen before: yellow rattle, properly known as *Rhinanthus minor*, which Andrew explained belonged to the near-unpronounceable plant family of the Scrophulariaceae. These are not particularly remarkable plants to look at. They grow to sixty centimetres or so tall, usually less, with a central spike carrying small, tubular yellow flowers. The delicate petals of each flower form a yellow tube with a purple tip and protrude from within a tough, papery green calyx (the fused sepals of the flower), which shelters and protects them. Most of the flowers had already gone over – the main flowering period is May and June – but a few were still fresh and we spotted a garden bumblebee visiting them, a very long-tongued bumblebee and one of the few British bees able to reach the nectar in rattle flowers. Once pollinated, the rattle petals wither and a seed pod swells inside the calyx, the flat, disc-like seeds hardening and becoming loose inside the pod. As we walked through the meadow the disturbance of our feet caused the stems to sway and the seeds to rattle pleasingly – the origin of the plant's name.

I had no more cause to think about yellow rattle until some years later, when I was investigating the declines of many UK bumblebee species. It had become clear that most of the bee species that had declined severely were the longer-tongued species and those that tended to emerge late from hibernation; species such as the great yellow, shrill carder, ruderal and short-haired bumblebees. When I began studying bumblebees in earnest I had seen none of these species and was desperately excited at the prospect. It also became clear that little was known about these bees. Searching the scientific literature revealed hundreds of studies of buff-tailed bumblebees and other common species, but almost nothing had been written about the rarer bees. I was determined to put this right – after all, we could not hope to conserve these creatures if we did not know anything about them.

By the time my interests in bumblebee conservation really came to the fore I had moved from Oxford to Southampton University, and I travelled far and wide in the UK to track down these exotic-sounding creatures. I saw my first brown-banded carder at Dungeness in Kent; red-shanked carders and ruderal bumblebees on Salisbury Plain; shrill carders on the Somerset Levels; and great yellows on South Uist. I went abroad, to Russia, Poland, Switzerland and even New Zealand, to find healthy populations of the rare species. Together with my students, I amassed data on the food plants used by all the different species that we saw, both common and rare, eventually accruing tens of thousands of records. We wanted to know how the various bee species differed with regard to the flowers they preferred, and whether the rare bumblebees were dependent on particular, rare flowers. In all this work, some clear patterns emerged. The sites that supported lots of bumblebee species, and those where the rare bumblebees were to be found, all tended to be flower-rich grasslands. There was usually a lot of red clover, which long-tongued bees favour as a source of pollen, and often there was yellow rattle.

While the queens of our common bumblebee species emerge early in the spring, sometimes as early as February, the queens of our rare species tend to emerge much later. Shrill carders and great yellows are not seen much before June, having slept since the previous September. This may be in part because the flower-rich grasslands they inhabit actually have few flowers in spring, despite their name. Grassland flowers tend to blossom from late May onwards, so it makes little sense for the bees that live there to wake up any earlier; if they did, they would starve. Yellow rattle is one of the first grassland flowers to bloom, and its deep flowers provide vital nectar and pollen for the hungry queens coming out of hibernation. It thus makes sense that the places where these bumblebees thrive tend to have good strands of yellow rattle.

It turns out that there is much more to the relationship between bumblebees and rattle than this. Since I was taught plant identification by Andrew Lack all those years ago, plant taxonomists have moved yellow rattle from the Scrophulariaceae to the no-more-pronounceable Orobanchaceae.* The common name of this family is the broomrapes, and therein lies the clue to their unusual biology. Some members of the broomrape family are

---

* Plant taxonomists seem to be a perverse bunch. Not only do they insist on family names that have at least a dozen or more syllables, but they then change them every five minutes, so that whatever names you manage to learn are soon out of date.

parasites on other plants, particularly legumes such as beans and broom. They don't bother producing green chlorophyll to photo-synthesise, but instead send out long roots that latch on to the roots of other plants and drain their nutrients. The bulk of the plant is below ground, for it has no need of sunlight, but in spring each plant sends up a yellow, sickly-looking flower spike which, despite its anaemic appearance, is quite attractive to bees. Broomrapes are rare in Britain, but they are common in more southerly parts of Europe and can even be a pest of broad-bean crops in the south of Spain. Yellow rattle is a cousin of the broom-rapes, and although it has green leaves and photosynthesises like a normal plant, it does share some of their parasitic habits. The roots of rattle latch on to those of nearby grasses, drawing out sugary sap. Patches of rattle within a meadow are often conspicuous, for the surrounding grasses are stunted and yellowed, weakened by the parasitism – reminiscent of vampire victims that are being slowly drained of life.

Scientists at the UK's Centre for Ecology and Hydrology in Wallingford have explored whether rattle might not be a powerful tool in the conservationist's armoury. The recent realisation that we have destroyed most of Europe's flower-rich grasslands has led to efforts to restore damaged meadows, or even to create them from scratch, as I have been doing at Chez Nauche. However, the great enemy of such projects is high soil fertility. Beautiful, diverse grassland that took centuries to develop can be reduced to a green sward of grass with barely a flower in sight within just a year or two, by the single act of adding chemical fertiliser. Ancient grass-lands tend to have low soil fertility, so grasses grow slowly and there is lots of room for other plants. In these conditions plants that are able to extract nitrogen from the air are at a huge advan-tage, and so legumes tend to thrive, for they have root nodules containing nitrogen-fixing bacteria. Hence flower-rich grasslands are rich with trefoil, vetches, meddicks, restharrows and clovers – all legumes. This is also good news for bees, since they love the pollen and nectar of legumes. Chuck on a sack of fertiliser and the grasses sprout up like crazy and smother everything else.

Once fertiliser has been added to a meadow, it is exceedingly difficult to remove it. The fertility will slowly decline if no more fertilisers are added and if a hay crop is removed every year, but this can take decades. By attacking grasses and reducing their ability to dominate the sward, yellow rattle offers the possibility of speeding up the process. To test this idea, Richard Pywell and his colleagues at the Centre for Ecology and Hydrology set up an experiment at Little Wittenham Nature Reserve in Oxfordshire, a cluster of pretty, steep-sided chalky hummocks just south of the River Thames. By coincidence, I used to go sledging there in winter when I was a PhD student at Oxford. Pywell sowed patches of species-poor, 'improved' grassland with yellow rattle seeds, after first scarifying the surface to provide some bare ground. The rattle established well, and quickly began to suppress the grasses. After two years he sowed a mix of wild-flower seed into the experimental plots, and then monitored whether they established successfully. The results were quite clear: the wild-flower seed mix took much better in plots with lots of rattle, presumably because there was less competition from the grasses.

It seemed to me that my meadow provided an opportunity to follow this up, and to simultaneously boost its floral diversity. Rattle has many hemiparasitic relatives; there are other species of rattle, such as greater yellow rattle and narrow-leaved rattle, and also eyebright, red bartsia, meadow cow-wheat, and so on. These other species might also prove effective at suppressing grasses and boosting floral diversity; perhaps some might even be better than yellow rattle? So in September 2010 I returned to Chez Nauche with a group of volunteers, my PhD students Andreia Penado and

Leanne Casey, and two staff from the Bumblebee Conservation Trust, Pippa Rayner and Tasha Rolph. At the time I had a very old Yardman sit-on mower, and I set about mowing 120 ten-by-ten-metre plots; the mowing was needed to get rid of the vegetation that had sprung back since the July hay cut, and to provide some bare ground for the seeds to germinate in. Unfortunately my Yardman expired after just a handful of plots, and I have never managed to coax it back into life since. Not having the funds to replace it with a new one, I rushed out and bought a normal lawnmower, and spent the next six days mowing neat squares into the field. The team of four girls followed behind me, sprinkling in seeds and performing a strange, collective shuffling dance to bed the seeds in (Pywell's team used a tractor-mounted roller, a far more sensible approach).

It was not long before my farming neighbour, Monsieur Fontaneau, happened to drive along the track adjoining the meadow, with the larger of his two sons in the passenger seat. He stopped, and they stared for a while. They turned their engine off and wound down the window. I waved, and he waved back. They stayed for perhaps ten minutes, before they drove away.

Half an hour later another car came along the track; it doesn't go anywhere, petering out just a kilometre beyond the meadow, so they were unlikely to be random passers-by. They too stopped and watched for a while, and then reversed away. Over the next few days perhaps a dozen different cars came by, roughly a dozen more than would normally pass the meadow. Clearly word had got out that the Englishman was up to something strange. I think I recognised the sturdy lady who had nearly shot me seven years earlier. Monsieur Fontaneau came back with his thin son. They all stopped and watched for a while, though none ever came over to ask what we were doing, and I didn't attempt to approach them with an explanation, which I felt sure I would be unable to provide in my hopeless French. To this day I wonder what they thought we were doing, mowing endless neat squares in a vast field, then performing strange, ritual dances in them.

I've been back to Chez Nauche every spring since to survey the plots, counting and identifying every plant species present. It takes several days, and is repetitive but rewarding work. Identifying some of the grasses is a particular headache, but I enjoy the challenge. The character of the meadow changes from year to year, depending on spring rainfall and temperature. Sometimes it is like a jungle, with the taller plants reaching head-height, while in drier years it is much sparser with patches of bare soil. Every year we find new plant species that I have not seen before. The yellow rattle has certainly established itself, though it is very patchy, and there is also some greater yellow rattle and small scatterings of the other species. Clovers seem to be spreading rapidly in many of the plots, so that they are thick with bees. Shrill carder queens, which love red clover, have become common in late May.

It is too early to say yet whether the different hemiparasites are having any effect. I have long since forgotten which plot is which (although of course I have this written down and filed away) – a good thing because otherwise I might subconsciously bias the results. In a year or two I will analyse them and see what it shows. Whether the experiment produces positive results or not, it is clear that the meadow as a whole is slowly improving, with a few more flowers every year. I do not know whether it will ever reach the dizzy heights of floristic diversity found in an ancient hay meadow, but it seems to be turning into a fair facsimile.

One of the few places in western Europe where there are still quite a few pristine flower-rich meadows is in the Alps. The sheer inaccessibility of many of the higher pastures has protected them to some extent from the ravages of modern agriculture, and the Alps remain the largest biodiversity 'hotspot' in Europe. A visit to these high meadows in late spring or summer is a must for any nature-lover. The weather is hugely unpredictable, but if you are lucky and catch the Alps on a sunny day they are breathtaking. Lush carpets of flowers tumble over the rocky slopes, buzzing with insect life making the most of the short summer. Of course the backdrop of dizzying snow-capped peaks and deep valleys with glimpses of distant lakes sets everything off nicely, too. Because of this extraordinary beauty and diversity, the staff at the University of Stirling decided that Switzerland would be an ideal place for a summer field course for our biology students. Despite the high cost of living in Switzerland, it was surprisingly cheap to rent a huge chalet high in the mountains, presumably because many of these buildings lie empty through the summer, the bigger influx of tourists coming during the winter to ski. To my delight, when we arrived at the chalet for the first time in June 2009, the first plant I noticed was yellow rattle, growing around the edges of the rough lawn. It turned out to be common in the area, not so much in the very high meadows, which tend to be heavily grazed by cattle, but in the lower areas downhill from our chalet, where the fields are kept free of livestock in the summer to provide a hay crop.

We set the students to work on a range of projects, one of which I supervised, on the diet of Swiss bumblebees (no muesli or chocolate for them). The Alps have all sorts of bumblebee species, including almost all the British species and a whole heap more besides, and identifying them was initially a challenge. I was particularly excited to see great yellow bumblebees, in Britain a rarity largely confined to coastal grasslands of the Hebridean islands and seldom found more than a few metres above sea level, but here happily living a kilometre or two higher amongst the mountain peaks. Despite the lack of bilberries there were plentiful bilberry bumblebees, among the most colourful species, with huge red bottoms and yellow stripes. On our first morning we recorded sixteen different species within a short walk of the chalet.

Amongst these was a species I had seen only once before, in the mountains of southern Poland – a species with no English name, *Bombus wurflenii*. It is one of a very small number of bumblebee species that has become a specialist nectar-robber, a professional thief, which makes its living by cutting holes into the side or back of flowers and stealing their nectar. Because it does not enter flowers by the conventional route it does not contact the reproductive parts and so does not pick up or transfer pollen. If flowers could speak (or think), one might imagine them shouting, 'Oi, stop! Thief!' Britain has species such as the buff-tailed and white-tailed bumblebees, which sometimes stoop to robbery if they are hungry and encounter a flower that is too deep for their short tongues – they are opportunistic thieves. In contrast *Bombus wurflenii* are incorrigible larcenists. They are well equipped for their profession. The mandibles of most bumblebee species are approximately smooth and paddle-shaped, well adapted for moulding wax and feeding broods. Buff-tailed and white-tailed bumblebees have small teeth on the edges of their mandibles, which help them to bite through the back of flowers when the need arises, but it takes them a little while to do so. In contrast the mandibles of *wurflenii* are armed with a row of long, curved and sharply pointed teeth, which enable it to slice though the sepals of most flowers in moments.

In the high Alps there are many deep, nectar-rich flowers, such as kidney vetch and monkshood, that aim to attract long-tongued bumblebees such as garden bumblebees and great yellows, and it is these that *wurflenii* targets. We also noticed that they were visiting, and robbing, yellow rattle. I'd never seen yellow rattle robbed before; the tough sepals are enough to deter amateur thieves such as buff-tails, so the rattle in my French meadow remains unburgled, but here in the Alps almost every flower had been robbed. Just as a house break-in is given away by a shattered window pane or a splintered door, so nectar robbery leaves behind obvious marks; *wurflenii* create crescent-moon-shaped holes in the sides of the sepals. The holes stain brown around the edges, and so contrast with the fresh green of the undamaged sepals. There is a certain irony that rattle, itself a parasite of grasses, is in turn parasitised by bees (which are parasitised by other bees, which are parasitised by mites, and protozoans, and so on and so on).

In the second week of our stay the students were assigned different projects, working in smaller groups, and one of the projects that we devised was to look at whether the robbing of rattle had a detrimental effect on seed set. The robbing holes remain visible long after the petals have wilted and the flower has set seed, and so for the older flowers on the plant – those low down on the stems – it was possible to count the seeds of both robbed and unrobbed flowers and compare them. If, by stealing nectar, robbing made the flowers less attractive to long-tongued, pollinating bumblebees, then we would predict fewer seeds in the robbed flowers. The students set to work, carefully pulling apart hundreds of old flowers and counting the flat green seeds within.

After a while of this tedious work, one of them, a bearded, ginger-haired Scot named James Morrison, noticed something odd. Within the patch of rattle in which they were sitting, almost every flower had been robbed on the same side, the left. When he counted them up, 98 per cent of the holes were on the left-hand side, with just a very few on the right. He mentioned this to us and, once it was pointed out, the pattern became obvious. What is more, when we looked at other patches of rattle in different meadows nearby, almost all showed a strong pattern, but in some meadows all the flowers were robbed on the right, while in others they were robbed on the left. Meadows just a few hundred metres apart showed opposing patterns.

Many years earlier I had carried out a brief study of 'handedness' in bumblebees. Some flowers have lots of florets arranged in rings around a central stem – clovers, for example – and I'd noticed that some bees always seemed to go clockwise around each flower, while others went anticlockwise. I persuaded one of my PhD students, Andrea Kells, to spend a week following as many bees as she could and recording which way around flowers they went. She found that every bee seemed to have a preferred direction of rotation, although they did occasionally switch, but that there was no overall tendency within any particular bumblebee species to be either clockwise or anticlockwise bees. When I did a little digging in the literature I found all sorts of strange examples of similar behaviour, often called 'handedness' despite the creatures in question having a distinct lack of hands. Some of the more peculiar studies I unearthed involved creeping up behind animals and surprising them. When startled by something approaching them from the rear, apparently both goats and snakes tend to turn round quickly, with each individual always tending to turn the same way – either left or right. We published our work in a fairly obscure journal and it sank without trace, attracting no interest or attention whatsoever from other scientists.

The 'sidedness' of robbing in rattle reminded me of this long-forgotten work. It was easy to imagine that *wurflenii* might have individual bias or handedness, but it was hard to see how that would result in every flower in a patch being robbed on the same side, unless a single bee had robbed all of them. Some of the patches were huge, with many thousands of flowers, so this just wasn't possible. We were intrigued, and so for the next three summers, whenever we had a spare moment from teaching the students, one or other of the staff would sneak off to find patches of rattle and score the number of robbing holes on the left and the right. We also watched the bees, recording whether *wurflenii* tended to land on the right or left of a flower. We noticed white-tailed bumblebees acting as secondary robbers, using the holes sliced open by *wurflenii*, so we followed them too, recording the behaviour of each bee as it visited successive flowers.

Year after year we found a strong bias in every patch, but no geographic pattern and no consistency across the years. A rattle patch might be robbed almost entirely on the left side one year, and on the right the next year. We found that pretty much every bee in a patch, both those robbing *wurflenii* and the secondary robbing white-tails, tended to approach flowers on the same side, be it left or right, and that this matched the location of the holes. So far as we could discern there was only one plausible explanation: the bees within each patch were copying each other. This might sound a little far-fetched, but actually it has been suspected for a very long time that bees might observe and learn from each other's behaviour. In a letter to the *Gardeners' Chronicle* written in 1857, Charles Darwin observed:

> One day I saw for the first time several large humble-bees visiting my rows of the tall scarlet Kidney Bean; they were not sucking at the mouth of the flower, but cutting holes through the calyx, and thus extracting the nectar . . . The very next day after the humble-bees had cut the holes, every single hive bee, without exception, instead of alighting on the left wing-petal, flew straight to the calyx and sucked through the cut hole . . . I am strongly inclined to believe that the hive-bees saw the humble-bees at work, and well understanding what they were at, rationally took immediate advantage of the shorter path thus made to the nectar.

More than 150 years later Ellouise Leadbeater at Queen Mary, University of London, showed that buff-tailed bumblebee workers that encounter robbed flowers were more likely to become robbers themselves, so that the behaviour spread rapidly from one individual to many. What we seemed to have discovered was that bees could not only learn from others to rob flowers, but that they copied the particular technique – robbing on the left or the right.

Our interpretation of the patterns of robbing found in our rattle patches in the Alps is as follows. In late May the first few flowers come into bloom. It probably takes a day or two before any bees discover them, for of course each year the worker bees have never seen rattle before, as none of them live for more than a few weeks. This is true of any flower – when it comes into bloom it takes a little while before any bees realise that it is worth visiting. Once one bee discovers it, others copy. The first *wurflenii* to discover rattle in a particular patch presumably has its own inherent bias, tending to land and rob on the left or the right, and if other bees in a patch then copy her behaviour, they will all end up being either left-handed or right-handed. We never had the opportunity to visit Switzerland earlier, in this formative stage of the process, because we were always occupied with end-of-semester exams at the time, but we were able to study it indirectly by examining the robbing patterns on flowers of different ages. The lowest flowers on a rattle stem open first, and the highest ones last, so that a single stem may be in flower for six weeks or more. Hence each stem carries a record as to what has been happening from when the very first flowers opened. We scored robbing of each flower from the lowest to the highest, and discovered that the patterns of robbing were less biased to one side or the other on the oldest (i.e. the first) flowers. One might imagine that sometimes more than one *wurflenii* might discover a patch at the same time, and if they have different strategies this would create a mix of holes on both left and right. This is not very efficient for the bees, for it is quicker to rob a flower using an existing hole than to cut a new one. Somehow over the next few days the bees seemed to settle on a common strategy, for flowers a little higher up the stem become uniformly robbed on one side or the other. Presumably newcomer bees benefit from adopting whatever is the most common strategy in the patch, which then makes the strategy even more common.

This may all seem a little obscure – why does it matter where the robbing holes on rattle flowers are? It presumably makes no difference one way or the other to the plant, which has its nectar stolen either way. Of course in the grand scheme of things this probably doesn't matter much, if at all, but it is a nice example of the ability of insects to adapt and learn from one another, and it is this ability that lies at the heart of the success of social insects.

In fact, as far as the rattle is concerned, it seems not to matter that it is robbed at all – James and his fellow students could find no difference in the seed set of flowers that were robbed compared to those that were overlooked by the robbers. This seems counter-intuitive; after all, if they are having their nectar stolen, and the nectar is there to attract pollinators, surely this should reduce visits by the long-tongued bees – the legitimate pollinators – and thus reduce seed set? It turns out that it is not that simple. Many studies have been carried out on the impact of nectar robbery on a broad variety of flowers, and while some studies found that robbed flowers set fewer seeds, other studies did not, and a few even found that robbing increased seed set. The explanation is not clear. Sometimes nectar-robbers do touch the anthers and transport pollen, so although they are robbing they are also providing pollination by accident, as it were (of course almost all pollination is by accident, since few insects set out to pollinate flowers). It may also be that the low nectar levels that result from regular robbery make the long-tongued bees work harder to get enough food, forcing them to visit and pollinate more flowers. Of course if that were true, then flowers growing in areas without nectar-robbers would benefit from producing less nectar, both saving on nectar and getting better pollination.

I am sure there is more to be discovered about rattle and its complex and unpredictable web of interactions with other plants and animals. James's discovery of the sidedness of robbery in the Swiss mountains illustrates that almost anyone can discover something interesting and new, if they simply take the time to observe nature carefully. He needed no special equipment, just curiosity and a willingness to look and think. His observations may not have changed the world, but they did lead to some interesting research that has expanded our understanding of the natural world by a tiny increment, leaving us just a little wiser than we were before.

CHAPTER TWELVE
Smutty Campions

1 September 2012. Run: 40 mins 40 secs. A lovely late-summer morning, with just a few puffs of white cloud in the sky. People: none. Dogs: 6, including that darned spaniel again, which managed to pull off my shoe in a surprise attack. Butterflies: 8 species, including lots of nymphalids – tortoiseshells, peacocks and red admirals – stocking up on buddleia nectar for the long winter ahead. There were also heaps of tired male bumblebees lolling around on the knapweed flowers, their fur bleached white by the sun – their days are numbered, their purpose done, for the new queens are by now all mated and safe in hibernation.

Most people have never even considered the question 'Why do males exist?' We simply accept that there are two sexes, in roughly equal numbers, and that offspring are produced by sexual reproduction between a male and a female. Whilst most of us are content, perhaps even pleased, that this is the way things are, evolutionary biologists have been fiercely debating the explanation for the predominance of sexual reproduction in both animals and plants for fifty years or more. If females simply produced copies of themselves (reproduced asexually), life would be much simpler and there would be no need for males at all. There would be no need for all the angst and aggravation involved in finding and courting a mate; no males fighting over access to females; no sexually transmitted diseases; no jealousy, cuckoldry or unrequited love. A female who simply copied herself could quickly out-reproduce sexual competitors, for all of her offspring would be reproductive females, rather than half of them being males. Indeed, some animals and plants do use this strategy; they don't bother with sex. Bdelloid rotifers (obscure microscopic animals found in ponds) eschew sex altogether, and seem to have survived perfectly well for many millions of years. Similarly some types of stick insect (including the Indian stick insects so often kept in schools), some nematode worms and even one or two species of lizard exist only as females. For most of the spring and summer, female aphids rapidly clone themselves so that one can become hundreds in a few weeks. So why is this not the norm?

Of course life might be rather boring in an asexual world, for there would be no peacocks' tails to impress, no magnificent antlers with which to battle rivals, no beautiful flowers to attract bees. And, of course, no sex. Perhaps fortunately, animals such as the stick insect are in a small minority; the vast majority of animals and plants prefer to have sex in one form or another. It seems that sex confers advantages, although they are hard to pin down precisely. The consensus is that the benefit derives from the mixing of genes from generation to generation, so that endless possible combinations are produced and no two individuals are exactly the same. This makes it possible for favourable combinations of genes to come together, enabling much faster adaptation. It also makes it harder for parasites and diseases to spread, because their hosts are all slightly different from one another, varying in their susceptibility to each disease. It seems that these barely tangible and much-debated advantages outweigh the cost and bother involved in sexual reproduction.

In most animals, individuals are either male or female (of course one might also question why there are two sexes, not three, or four, or fifty-seven, but let's not go there just yet). In sexual reproduction each sex produces gametes – sperm from males, and eggs from females – and these must come together to form offspring. In contrast, most plants are hermaphrodites: each plant has both male and female sexual organs. More often than not these are contained within the same flower, and they are usually easy to see. The male parts, the anthers, are usually supported on thin stalks and produce powdery pollen, often bright yellow in colour, although it can be white, orange, purple or scarlet. The female parts are often white, generally involving a long stalk with a sticky tip (the stigma) to trap pollen.

Having both male and female parts in the same flower has an obvious disadvantage: it is all too easy for plants to accidentally mate with themselves, which is rarely a good idea except as a last resort. Plants have evolved many mechanisms to avoid this, the simplest of which is to have flowers that are male and female at different times. If you have geraniums* in your garden, take a close look at the flowers. The young flowers are male, each with ten stamens producing pollen, which is often a distinctive purple colour. After a day or so the stamens wither and the female part ripens, forming a central pillar to the flower, which unfurls into a beautiful five-pointed star, sticky to trap pollen. The male and female versions of the same flower cannot mate with one another because they are separated in time. In foxgloves the flowers hang from a

---

* Geraniums should not be confused with pelargoniums, the red-flowered stalwart of every hanging basket, which are often mistakenly called geraniums. True geraniums include many native, wild species, such as herb robert and the lovely purple-flowered meadow cranesbill (so called because the seedhead resembles the head of a bird with a long bill, one of the distinguishing features of the geranium family). There are also many perennial herbaceous garden varieties, most of which are good plants to encourage bumblebees.

tall stem, with the oldest (female) flowers at the bottom and the youngest (male) flowers at the top, exploiting the fact that bees have the habit of starting at the bottom and working upwards. An arriving bee deposits any pollen she is carrying from another plant on to the stigmas of the lower female flowers before picking up a fresh pollen load from the upper male flowers, which she then carries off to another plant.

Few plants follow the animal model of having distinct sexes, but this small number includes one of the more common flowers in my field in France, the white campion. Campions are distinctive and hence easily recognised plants with five petals, each with a notch or cleft at the margin that is slightly reminiscent of Kirk Douglas's chin. The flowers of the male and female plants are readily distinguished. The males produce many stamens and yellow pollen that contrasts with the white petals, while the females produce a cluster of twisted, entwined white stigmas. They make attractive garden plants; in my garden in Dunblane I had both white campion and its relatives red campion and ragged robin, the latter with naturally tattered petals. White and red campions flower from May onwards; the female plants are quickly pollinated and pour their energy into producing chunky capsules filled with large, round seeds, so that by the end of June there are few female flowers. In contrast the male plants continue to flower right through the summer until October, a seemingly futile effort, for there is precious little chance of their pollen successfully fertilising a female. The gardener can use this difference between males and females to advantage; by weeding out all but a few female plants, you can ensure far more flowers for longer.

Campions are pollinated by both moths and bees in the familiar mutualistic relationship, providing nectar to attract and reward their pollinators. However, there is an unusual twist, for some of the moths that pollinate them then lay their eggs upon the flowers, gluing them firmly to the petals. One such moth, appropriately named 'the campion', feeds exclusively as a caterpillar upon the developing eggs of the female flowers. The small caterpillars live inside the capsule where they are safe from predators, their presence usually given away only by a tiny hole from which protrudes a cluster of droppings, or frass as it is known to us entomologists. Eventually the caterpillars grow too large to reside entirely inside the capsule and bite a larger hole from which their camouflaged but fat rear end protrudes, their head still munching away on the remaining seeds inside. Sometimes, when moth populations are high, almost all of the campion seeds in my field at Chez Nauche are consumed, so it is fortunate that campions are perennial plants that will get further chances to produce seeds in subsequent years.

For campions there is a second risk to sexual reproduction, aside from the danger of attracting egg-eating moths. Campions suffer from one of the most virulent of sexually transmitted diseases of plants, the campion smut *Microbotryum violaceum*. Smuts are parasitic fungi that are often spread by pollinators and so, in the case of campions, pollinators pose the double threat of both laying eggs that will hatch into voracious larvae and giving the flower a nasty dose of the clap. Smuts have a complex and fascinating life cycle. When fungal spores are delivered by a moth or other pollinator to a campion flower they germinate, sending out thread-like fungal hyphae, which invade the plant tissues. The fungus spreads through the plant, slowly invading right down into the roots. It produces sporidia – the fungal equivalent of sex cells – which are free-living within the host plant. If the plant is infected by more than one 'mating type' of smut, then the sporidia from the two different types can fuse: the fungus version of sex. Since there are many different 'mating types', and these sex cells are not differentiated into the equivalent of eggs or sperm, there are effectively dozens of different sexes in fungi.

Whether it is infected with one or many mating types of fungus, the plant is doomed to a life of sexual slavery. It seems that they rarely, if ever, throw off their infection. The fungus hijacks the flowers of the plant to its own ends. In a male plant the fungal hyphae invade the anthers so that, instead of producing yellow pollen, they produce fungal spores. These are dark purple in colour, and burst from the anthers even before the flower bud opens, so that when the petals unfurl they are stained purple. It is an ugly sight – the pristine white petals sullied by blotches of purple, which smear like running mascara with rain or dew. In the first year of infection usually just a few flowers are invaded, but by the following year they are all dark with fungal spores, and the plant has lost any chance of reproducing.

When it finds itself in a female campion, the fungus has a further trick up its sleeve. It subverts the development of the female flowers, preventing development of the ovaries and forcing the plant to produce fake anthers, once again bursting with fungal spores. Once they are infected, male and female plants are rather hard to distinguish, although if you pull the flowers apart, the stunted ovaries of the female flowers are still visible.

In some years a great many of the white campion plants at Chez Nauche are infected with smut, to the extent that it can be quite hard to find any healthy flowers. How the species survives in such abundance is something of a mystery to me, for the majority of the population have been effectively sterilised, and any uninfected plant is likely to be exposed to fungal spores very swiftly. Presumably some individual plants have resistance, and the genetic mixing involved in sexual reproduction helps to keep at least some plants one step ahead of the parasite.

I started this chapter with the question 'Why do males exist?' A related and equally good question that one might ask, but would probably never think to do so, is 'Why are there usually roughly equal numbers of males and females?' This is generally true of most animals, and also of plants such as campions in which the sexes are separate. But why? We might accept that sex is useful in allowing genes to be regularly jumbled up, giving scope for speedier evolutionary change, but why do there have to be so many of us males? In most animals, males invest very little in reproduction – sperm are cheap, and one male could cheerfully inseminate numerous females. In red deer, for example, the dominant males defend a huge group of females during the rut and thereby gain exclusive rights to mate with them. Most males aren't strong enough to compete and so don't get to mate at all, living out their sad, frustrated lives on the edge of deer society, consumed with jealousy for the monarch of the glen (okay, I'm getting a little carried away now, but you get the idea). So why are so many males produced? Why not just have enough males to go round, and lots more females to produce and rear offspring? A population could certainly grow more rapidly with this arrangement. In fact in fruit flies, females produce more offspring when there aren't many males about, simply because the males constantly harass them and force them into mating repeatedly, when the females would rather be doing something more useful. The answer was provided by the evolutionary biologist Ronald A. Fisher as long ago as 1930, based at the time at Rothamsted Experimental Station in Hertfordshire. Fisher was a collaborator of E.B. Ford and one of the founders of modern statistics (something that not everyone would thank him for; he invented tests such as Analysis of Variance and, of course, Fisher's Exact test, which I recall being forced to compute laboriously by hand during A-level maths classes). He suffered from extremely poor eyesight, something that may well have saved his life, for it meant that he was rejected when he enthusiastically and repeatedly attempted to join the army at the outbreak of the First World War.

Fisher's explanation was remarkably simple. Suppose males are scarce in a particular population. Newborn males in such a population would have better prospects than newborn females, for they would be likely to have more offspring – there would be lots of mates for them. Therefore any parents with genetic tendencies to produce sons would be at an advantage, and their genes would spread. Males would become more common, but as the ratio of males to females approached 1:1, their advantage would disappear, and the genes favouring male production would cease to spread. Exactly the same argument in reverse could be used when starting with a male-biased population, in which the production of daughters would be favoured.

This argument is now known as 'Fisher's principle', and campions provide a rare and interesting exception. In white campions, females are generally more common than males. According to Fisher's argument, these females should have lower reproductive success than the few lucky males, who benefit from being able to pollinate the many females. Hence natural selection ought to favour individuals to produce more male offspring. Why doesn't this happen? The answer is complicated, but sufficiently fascinating to be worth explaining. So far as it is understood, the odd sex ratio in campions seems to be the result of a battle that rages between different genes within the plants. Oxford biologist and prominent atheist Richard Dawkins has championed the view that the genes within our body operate on an 'every gene for himself' basis. On the whole it is generally in the interests of genes to collaborate, for their chances of propagation into the next generation depend on their host body thriving and reproducing. Sometimes, however, individual 'selfish' genes attempt to cheat on this alliance, ensuring their own propagation at the expense of their collaborators.

Most of the genes within the cells of organisms such as humans or campions are packed into chromosomes within the nucleus. A few live outside the nucleus in the cell cytoplasm; there are genes in our mitochondria and in the chloroplasts of plants left over from an ancient time when these cell structures were actually free-living organisms in their own right. These genes are passed down through the maternal line (in eggs, but not in sperm or pollen), so if they find themselves in a male they have reached a dead-end. To reduce the chances of this happening, some cytoplasmic genes distort the ratio of offspring produced by their host, killing male offspring so that more female offspring can be produced. This is bad news for the rest of the genome, for it will find itself within females in a population in which males are scarce and hence have by far the higher reproductive success. It is particularly bad news for genes on the Y chromosome, for they are found only in males and so bear the brunt of this genetic revolt (note that both campions and humans share a system in which females have two X chromosomes and males an X and a Y).

Genes on the Y chromosome don't take this lying down; the attack of the cytoplasmic genes selects for genes on the Y chromosome that can inactivate the male-killing cytoplasmic genes – so-called 'restorer genes'. In turn the cytoplasmic genes evolve to evade the blocking action. This battle is ongoing, and so the relative abundance of male and female campions in any particular population is a reflection of the state of the endless war; if the cytoplasmic genes currently have the edge, the females are disproportionately numerous, whereas if the Y-linked genes have successfully knocked them out and gained the upper hand, the ratio is 1:1.

You may well be feeling that this is quite enough genetics for one day, but there is more, for white campions do not operate in genetic isolation; they hybridise with their cousin the red campion, and this has intriguing consequences for the genetic war within.

In the UK white campions are more common in the drier, sunnier and more arable east of the country, often occurring in field margins and fallow ground, while red campions are more common in the wetter west; they abound in the banked hedgerows and road verges of Devon and Cornwall. Red campions seem to be more shade-tolerant and damp-loving, often growing in woodland glades and edges, while white campions are usually found out in the open. In some areas, including Hampshire where I used to live, both species are common and frequently occur within close proximity to one another, and in such places pink-flowered hybrids are common. During my field work on bumblebees, travelling around south-central England, I noticed many clusters of pink-flowered campions, but also no shortage of both white and red parents. Ever the scientific dabbler, this struck me as interesting and worthy of study. If the two species can interbreed so easily, what keeps them distinct? Why don't they just blend together as a mass of pink-flowered campions? Moths and bees do not stick to woods or to meadows, but readily move between the two, so red and white campion females must often receive pollen from males of the other species. Clearly this resulted in offspring: the pink-flowered hybrids. Unless these hybrids were in some way less fit than their parents, and thus were selected out of the population, it was hard to understand how the two parent species maintained their integrity. It was possible that part of the answer might lie in the interaction between the cytoplasmic genes that favour female production and the Y-chromosome restorer genes, for in hybrids these will become separated; Y-linked restorers from one species are unlikely to be effective against cytoplasmic genes from a different species, so we might expect hybrid populations to be heavily female-biased. If there were no males, this could lead to hybrid populations crashing.

I decided to do some experiments. The first step was to check whether the pink hybrids were fully viable and healthy. Some species can interbreed, producing hybrid offspring (lions and tigers, for example), but the hybrids are sterile and so the two species cannot merge together. I crossed red and white campions in the glasshouses at Southampton University, keeping the males and females in separate sections of the glasshouses and pollinating them by hand using a paintbrush. I reared the pink-flowered offspring, which grew vigorously, and the following year I crossed the hybrids with one another and with both parent species. Once again all the offspring were healthy, and I soon ended up with a lovely array of flowers in all shades from white to red. If anything, the hybrids were larger and healthier than their pure-blood parents, which left unanswered the question of how the parent species remain distinct. Although there were more females than males in the hybrid offspring, there still seemed to be plenty of males to go round.

It occurred to me that these hybrids might be less healthy if they were growing in a natural situation, where they would be exposed to competition from other plants and to grazing by herbivores, rather than in the benign environment of the glasshouses. Perhaps the hybrids would be ravaged by slugs, or all their seeds would be consumed by campion moths, and this could explain why the two parent species did not blend together. I decided to set up a long-term field experiment. At the time Southampton University was fortunate to own a substantial country estate, Chilworth Manor, just to the north of the city. The manor had been converted into a rather posh hotel and conference centre – in fact I had been put up there when I was interviewed for the lectureship at Southampton. The grounds included a traditional Victorian walled garden, which was available for members of the biology department to conduct experiments, and extensive woodlands. I planted out replicate patches of both white and red campions, and their hybrid offspring, in woodland and in the open, sunny walled garden. My prediction was that the white campions should thrive in the open, and the red campions in the woodland. If the hybrids failed to thrive as well as the parents in either habitat, because they were poorly adapted to either habitat or because the hybrid populations became dominated by females, this could explain their relative rarity in the wild.

I monitored these campion populations for nine years. In the woodland none did particularly well. Grazing by deer was heavy, and the white campions and the hybrids quickly died out, while only the red campions clung on, although they produced few flowers and fewer offspring. In the walled garden all of the campion patches flourished. Campion moths attacked many of the seed capsules of both white and pink campions, never the red, but there were always plenty left over and seedlings sprouted up everywhere. Pollinators moved between my patches, and soon most of the offspring were pink hybrids. The original parental plants eventually died, and after six or seven years there were few plants left that resembled the parent species. There was a female bias in the hybrid populations, presumably because the selfish cytoplasmic genes had become decoupled from the Y-chromosome restorer genes, but about 30 per cent of the populations were male, which seemed to be enough to pollinate the females. Overall – and contrary to my predictions – the hybrids seemed to be winning. I would have liked to continue this experiment for many decades, to discover the eventual fate of my campion populations, but it was not to be. To my frustration the university sold the walled garden to developers, and my campion patches were swept away in their ninth year, to make way for the shiny glass-and-steel buildings of a science park.

One of the great difficulties in ecological research is conducting long-term experiments. Grant funding rarely exceeds three years, so most experiments are designed to take no longer than this, yet any ecologist will tell you that many ecological processes occur over the timescale of decades or longer, and evolutionary processes usually take much longer still. Of course any large-scale experiment is also likely to need a fair bit of land, and land is cripplingly expensive in most of the UK. My campion experiment was not the greatest experiment ever, and perhaps I had found out all that I was going to, but its loss was annoying and was one of the driving factors behind my purchase of Chez Nauche. With my own piece of land I could set up experiments that could go on for as long as I liked, with no possibility that they would be destroyed by factors beyond my control (being in the middle of nowhere, there is little likelihood of a compulsory-purchase order for the construction of a bypass, housing estate or shiny new science park). So, in a roundabout way, perhaps I owe a small, begrudging debt of thanks to the university administrators who decided to flog off assets to raise money, and perhaps even to the politicians behind the university funding cuts that forced this decision, for without them perhaps my meadow in France would still be a cereal field.

For me, campions and their many associations are a lovely example of the interrelatedness of life. Genes flow within and between species, carried by moths and bees, along with fungal diseases that castrate male plants and turn female plants into false males. Moth caterpillars eat the seeds and are in turn predated by birds, or parasitised by tiny wasps or flies. If they survive to adulthood, the adult moths are hosts for tiny mites that live in their ears, and are food for the bats that hawk above the long grass, which are themselves hosts to fleas, lice, ticks and blood-sucking flies. Every living thing in the meadow – be it beautiful, mundane, gruesome or obscure – is linked to everything else, by just a few degrees of separation. The complexity of these myriad interactions is far beyond our ability to understand, and perhaps it is both arrogant and futile even to try. Maybe we should simply be happy that it is so, and try not to mess it all up too much.

PART III

Unravelling the Tapestry

I hope you realise by now that every creature has a story, and that most of those stories have yet to be told. There is so much that we do not know about almost all of the ten million or so species of organism with which we share the planet. It would be a terrible shame if these creatures were to be lost, be they elephants or earwigs, lions or ladybirds. What is more, these animals and plants, fungi, viruses and bacteria do not live in isolation. Their lives, and ours, are inextricably woven together. We don't know how it all works – and yet we are thoughtlessly picking it apart. Many species have been lost, and more disappear every day. The threads of the tapestry are being picked out, one by one. In this final section I will give you some glimpses of what we have done, and of the huge risks that we run by continuing to harm the environment on which we are utterly dependent.

I don't want to depress you. Nobody likes bad news, but please don't stop reading here. This is the most important bit. Life on Earth is wonderful and unbelievably complicated. It would be madness to continue to destroy it. It is not too late to ensure that our grandchildren inherit a world that is almost as rich in wonders as our own. The window of opportunity is open for us to act, although it will not remain so for much longer. But first we must understand where we are, and how we got here . . .

CHAPTER THIRTEEN

The Disappearing Bees

16 May 2013. Run: 40 mins 10 secs. A cool, misty morning, promising to brighten up. I'm feeling fuzzy-headed today; a pair of garden dormice kept me awake half the night with their angry chatter. As I set off I spotted a little owl perched in the dead elms. I thought I'd heard its distinctive hooting over the last few nights, competing with the racket from the mice. People: one farmer in his tractor in standard French farmer uniform of blue overalls. Dogs: 5. Butterfly species: 13, including a Glanville fritillary. The mist also helped me get close to a pair of frolicking hares, endearingly gangly and lolloping creatures.

To the north of my meadow in France the boundary is marked by a green lane, flanked by hedges. The hedge on my side is overgrown and about three metres thick. In spring it is full of blackthorn blossom and nesting birds, and in autumn it drips with fat purple sloes, perfect for sloe gin. I probably ought to cut it back, but I rather like it the way it is, and in any case I haven't saved up enough for a hedge-flail for my tractor. The hedge on the other side of the track is barely sixty centimetres tall and less in width. The farmer who owns the field beyond keeps it so heavily trimmed that it has little value for wildlife, and the stunted shrubs struggle to survive at all. He is an arable farmer, and sows his fields each year with wheat, maize or sunflowers. The many fields of sunflowers are one of the great pleasures of visiting the Charente in July. The serried rows of huge, nodding, sun-warmed flower heads, fringed with golden-yellow petals, are a sight to fill the heart with joy. Somehow fields of oilseed rape just can't cut it in comparison. Sunflowers are much loved by bees – both bumblebees and honeybees – and by other insects, and they flock to the cornucopia of nectar. Which would be great: gaudy flowers, happy bees. Except for one insidious, invisible factor that is perhaps one of the greatest threats to bees and to other farmland wildlife. But before I explain what that is, let me rewind a little.

Bees have been steadily declining for sixty years or more. Not just bumblebees, but probably most of the other wild solitary bees, too. I often say that these declines are undoubtedly driven by agricultural intensification, but that is a vague and woolly statement. Agricultural intensification has taken a myriad of forms: loss of hedgerows, abandonment of rotations, increased use of synthetic pesticides and fertilisers, mechanisation, drainage of marshes, and many more things besides. Our countryside has changed massively, but over such a long period it is very hard to notice. Just as one barely notices one's own children grow, so we struggle to detect gentle, steady changes. Until one day we look back and say, 'Hang on a minute, I can remember a time when I was a child when the buddleia bush in my parents' garden was covered in butterflies. Why hasn't mine got any?' But it is hard to be sure that it isn't just seeing life through rose-tinted spectacles. Was the buddleia bush really always covered in butterflies, or is it just that you remember one particular day when they were especially common? After all, wasn't it also always sunny?

For most of our wildlife it is hard to quantify how their populations have changed over the last fifty years. We can count bees, or earwigs, or voles (although each poses different challenges when it comes to devising repeatable, meaningful ways to count them), but we have few or no numbers from the past for comparison, so it is very difficult to say if there are fewer of them now than there were then. What we usually do have are distribution maps from the past (admittedly often patchy in coverage), and these can reveal that species have disappeared from parts of their former range. Among the bumblebees there are some dramatic examples. In the 1950s the great yellow bumblebee used to be found throughout Britain, from Orkney to Cornwall, but now it is confined to Orkney, the Hebrides and the far north coast of mainland Scotland.* The short-haired bumblebee was once widespread in the south-east, but is now extinct there. However, most of our species have not shown massive range declines. We can still find common carder bumblebees or early bumblebees almost everywhere in Britain, but whether there are fewer of them than there used to be we cannot say with certainty, although I would guess this is so.

For just a few wildlife groups we do have really good long-term data sets on how their populations in the UK have changed. Keen amateurs have been recording numbers of birds, butterflies and moths for many decades in repeatable ways, providing figures that we can compare directly over time. For example, the butterfly-monitoring scheme was started in the 1970s. Volunteers walk a regular route every fortnight throughout the spring and summer, identifying and counting every butterfly they see. Some routes have now been walked regularly for well over thirty years, and

---

* Interestingly, corncrake distributions, both past and present, closely match those of the great yellow bumblebee. Corncrakes used to be found nesting in hay meadows and cereal fields all over the UK, but the loss of hay meadows removed much of their habitat, and the switch to early-maturing winter cereals means that many of their nests are destroyed by combine harvesters. They now cling on only in the remote, crofted corners of Scotland where farming has changed relatively little.

there are currently in excess of 800 routes being walked every two weeks, scattered all over the UK. This provides the best long-term data on any insect group in the world. Birders have been doing similar things for even longer, so for birds and butterflies we can quantify exactly what is happening. Sadly it is mostly bad news, particularly for the birds, butterflies and moths that live on farmland.

We would expect most wild animals and plants to have declined from the 1950s to the 1980s, for we know this was the period of great changes in farming, when little heed was given to the needs of wildlife, and increasing food production seemed to be the sole priority. The policies in place were brought in during the Second World War, when Britain was cut off from external food supplies, and maximising food production was therefore understandably of paramount importance. These policies catalysed a great era of change, during which we lost our hay meadows, chalk grassland, hedges and much else besides.

By the 1980s things were changing. Food surpluses in Europe – grain and butter 'mountains' and wine 'lakes' – led to policies to remove land from agricultural production, much of it going into set-aside, with farmers receiving payments for not growing crops. Set-aside land was left fallow, giving a little breathing space for wildlife.* There was a dawning realisation that we had been

---

* Actually, to start with, much of the set-aside was useless for wildlife as it was often treated with herbicides to prevent weeds from seeding, and land was only left fallow for short periods, giving little time for wild plants and animals to colonise before it was ploughed up. Later iterations of set-aside schemes were much improved, allowing for the long-term set-aside of areas. Sadly, just as many of these were becoming havens for wildlife, EU policy changed and in 2008 more or less all set-aside schemes were abandoned.

steadily obliterating farmland wildlife, and that our trajectory might not be either sustainable or desirable. It was becoming apparent that farmers need bees to pollinate crops, and ladybirds, lacewings, ground beetles, wasps and hoverflies to eat greenfly and other pests. Chemicals such as DDT, heralded as a wonder-cure against all pest insects when it first came on the market just after the war, had long since been banned, when it turned out that they persisted in the environment for years, accumulated in food chains and were building up in humans. Raptor populations are only now recovering from one of the unexpected side-effects of this chemical; it thinned their egg shells, so that the eggs usually cracked before hatching.

I was at university in the mid-1980s, and there I was taught that the gold-standard of crop protection was called Integrated Pest Management, or IPM. The basic principle was that farmers should monitor the numbers of pests on their crop, and should only deploy control agents if and when there was a problem. They should try to manage their land to maximise the numbers of natural enemies, boosting populations of predatory insects such as lacewings by providing them with healthy non-crop habitat, such as hedgerows or strips of permanent tussocky grass – places in which they could spend the winter and find food when there were no pests on the crop. Cultural controls, such as crop rotations, can greatly help to reduce pest numbers, as the crops favoured by a particular pest are constantly moving from field to field and are therefore harder for the pest to keep track of. We were taught that chemicals should only be used as a last resort; and then only compounds that rapidly break down should be used, so as to minimise the impact on beneficial creatures.

More broadly, it had become clear that farmland biodiversity was important, both ecologically and economically. Hedges are not just boundaries between fields, but are reservoirs for beneficial insects, and provide flowers and nesting sites for bees. Changes to the Common Agricultural Policy were introduced that enabled farmers to be paid to promote biodiversity. Schemes were introduced to pay farmers to replant hedges and copses, to sow strips of wild flowers, even to create special 'beetle banks' from tussocky grasses. By the 1990s little good habitat was being lost, and lots of money was being poured into protecting and promoting wildlife on farms. The conservationists had won, and could finally put their feet up and watch the flowers grow.

Depressingly, that isn't quite how it panned out. The benefits to wildlife simply don't seem to have materialised, despite the injection of billions of pounds of taxpayers' money – currently running at about £500 million per year. After thirty years or so of agri-environment schemes, the data suggest that most farmland birds, moths and butterflies remain on a downwards trajectory, with populations dwindling year-on-year. There really are fewer butterflies on the buddleia bush – it isn't just my imagination. So what went wrong?

Partly the answer may lie with failures in the design of agri-environment schemes. Many were introduced as best-guesses, without much evidence that they would work. The entry-level schemes, whereby farmers get small payments for a range of simple measures to protect wildlife, often don't differ in any significant way from what the farmers would be doing anyway without payments. Many farmers don't really know what the agri-environment schemes are meant to achieve. They may sign up to create a 'species-rich grassland', and receive payments, but when the wild-flower seeds don't germinate or the grassland gets invaded by docks and thistles, they are often unsure what to do. The paperwork involved in entering the schemes is dauntingly complicated, and becomes more so with each successive iteration of these schemes, which are ever-changing. Farmers with small farms in marginal areas – often the ones with most wildlife on their lands – cannot afford to spare the time to fill in the forms or cannot make sense of them. Large, rich farmers pay agents to do it for them. Entry to the higher-level schemes – those that are most likely to actually benefit wildlife – is competitive, as most money goes on the entry-level scheme, so farmers who do fill in the forms may be rejected and find that they have wasted their time. Overall it is fair to say that agri-environment schemes have not been an unqualified success, and they probably provide the European taxpayer with poor value for money. Nonetheless, they are surely better than nothing. So why are farmland wildlife populations still heading steadily downwards towards oblivion?

The short answer is that we don't know for sure. But some people think they do, and I'm starting to come round to the idea that they may be at least partly right. In October of 2006 honeybee keepers in the USA found that their bees were disappearing. Whole hives that seemed perfectly healthy one day were deserted the next, the adult bees having simply vanished. There were no corpses, and no clues as to what had happened. Various names were coined, with Marie Celeste Syndrome being the most apt, but Colony Collapse Disorder is the clumsy term that stuck, often abbreviated to CCD. It didn't just affect one or two hives, but hundreds of thousands. Some beekeepers lost most of their hives and went out of business. The almond farmers of California found it almost impossible to find beekeepers who could supply hives to pollinate their crop, and the cost of hiring hives went through the ceiling.

Beekeeping in North America is very different from beekeeping in Europe. Here, many beekeepers have just a handful of hives and beekeeping is generally not big business. All but the very largest-scale beekeepers have fewer than 100 hives. Many hives remain in the same place all year round, although some beekeepers may move their hives a few kilometres to help with pollination of a particular crop, or to gather heather honey from the nearby hills in late summer. In contrast, in North America the big beekeepers have thousands of hives. So many hives cannot be kept together in one place for long, as they would soon exhaust the local food supply, so they are stacked on huge trailers and transported around the continent from crop to crop. Farmers who need pollination pay for the service, so the bees go to California for the almond blossom in March, to Florida for the flowering of the citrus orchards in April, north to New York to pollinate apples in May, up to Vermont for the blueberries in June, and then back to Florida via a few weeks in the pumpkin patches of Pennsylvania. Every year a colony may travel 18,000 kilometres. One might imagine that the bees (and the beekeepers) are exhausted; this is certainly not a natural way of life for a bee.

For both the beekeepers and the farmers the bees are vital to their livelihoods, so CCD was a disaster. It caused widespread panic and an urgent hunt for the culprit and a cure. Beekeepers elsewhere in the world started to look for signs of CCD, and the following year there were reports of similar problems in Europe, although on a smaller scale. Sometimes the symptoms were slightly different, but the panic spread, and the media published dramatic tales of a worldwide scourge, which threatened the very survival of bees (they rarely mentioned that CCD is seemingly confined to honeybees, just one of thousands of bee species).

Seven years on, and millions of pounds worth of scientific research later, we are still not certain what the answer is. Many consider the *Varroa* mite to be the prime suspect. This parasitic mite has spread from Asia throughout the globe in recent years, accidentally transported by humans, and is certainly a major threat to honeybee health. The mite sucks the blood of adult bees and the developing brood, spreading viral diseases from bee to bee as it goes, and it is hard to control. However, *Varroa* was around for quite a while before CCD, so it cannot be as simple as that. Others blame mobile phones, claiming that the signals interfere with bee navigation, causing them to get lost. Interesting though this theory is, there is not a shred of evidence to support it, and mobile phones were also widespread long before the arrival of CCD. Others blame genetically modified crops, but again this doesn't seem to stand up to scrutiny. A more plausible theory is that the diet of honeybees has become very narrow. Naturally honeybees feed on a huge range of wild flowers through the year, but in intensive agricultural landscapes they may get most of their food from just a few different crops, and for weeks on end they may be feeding on a single crop. This issue is particularly acute in North America, where bees are transported from one intensive agricultural landscape to another throughout the year. Just as humans require a balanced diet, so it may be that feeding on just a handful of different foodstuffs does not supply honeybees with all of the nutrients they require. Imagine if you were forced to eat only Brussels sprouts in December, bacon in January and chocolate in February; you might end up feeling more than a little off-colour. Finally, many suspect that pesticides may be to blame, and this brings me back to my French neighbour's sunflowers.

In the mid-1990s a new class of insecticide was introduced. Known as neonicotinoids, they are synthetic variants of nicotine. They block open insect nerve-receptors, thus attacking the insect nervous system and brain, and are phenomenally toxic in tiny amounts. Of course insecticides wouldn't be much use if they weren't toxic to insects, so this might be regarded as a good thing. Neonics (they sound a little more friendly when abbreviated) have a major advantage over most of the insecticides that went before, in that they are systemic. They can be applied as a seed dressing before the crop is sown, and the germinating seedling absorbs the chemical, which spreads throughout the plant. Any herbivorous insect that eats any part of the crop dies. This is a wonderfully neat system. Previously insecticides had to be sprayed on to the crop from a tractor-mounted boom. Much of them landed on the soil, and in even a slight breeze they would blow in to the hedgerows. Only the parts of the crop that were directly coated with the spray were protected, so the lower leaves and roots would be vulnerable to herbivores. As the crop grew, further applications were needed on the new leaves. Overall, more chemical was needed to protect the crop, and the farmer had to spend both time and diesel applying it. Many of the chemicals used were pretty nasty – for example, the organophosphate insecticides were derivatives of nerve agents developed during the war to kill people – and so spraying them around posed a direct threat to the farm worker. All in all, it is easy to see why the neonics proved to be hugely popular, and they quickly became one of the most widely used classes of insecticide in the world. They now comprise about one-quarter of all insecticides used globally, and one type of neonic, known as imidacloprid, is the second most widely used agro-chemical (after the herbicide glyphosate). In the UK, agricultural use of it has risen steadily, reaching about 80 tonnes per year at the last count.

The systemic nature of neonics is both their great strength and, perhaps, their Achilles heel. They spread to all parts of the plant, and that inevitably includes the nectar and pollen. If the crop is visited by pollinating insects, then they consume small amounts of these chemicals. Not long after the introduction of neonics in the 1990s, French beekeepers started to claim that they were causing their honeybee colonies to die. Their campaigns led to partial bans on some types of neonics on some flowering crops, but to the concern of French beekeepers, this largely meant that farmers simply used different types of neonics. This controversy rumbled on for some years until an incident in Germany in 2008. A batch of maize seeds had been coated with an incorrect formulation of neonic. The chemical was not properly stuck to the seed, and when the seeds were drilled, much of the coating blew away as a fine powder. Hundreds of honeybee hives in the area were wiped out more or less instantly. There was an uproar, and neonics were banned in Germany pending an investigation, but the ban was subsequently rescinded when it became clear that the problem lay primarily with the incorrect formulation. Nonetheless, this incident raised awareness among beekeepers and environmentalists as to the acute toxicity of these compounds to insects such as bees, and prompted further investigations.

Of course the chemical manufacturers were aware that neonics would get into the pollen and nectar of crops from the start. Agrochemicals go through various safety tests before they are licensed for use, including an evaluation of their toxicity to bees and other beneficial insects. Typically, groups of lab animals are fed on varying doses of a chemical and then monitored to see if they expire. This enables calculation of the 'Lethal Dose 50%' or LD50 – the dose that causes half of the test animals to die. The LD50s for the various different neonics in honeybees are all very low, just four- or five-billionths of a gram for the common types such as imidacloprid and clothianidin. To put that in context, one gram (not much more than the contents of a sachet of salt) is enough to give an LD50 to 250 million honeybees, or about twenty-five tonnes of bees.

How does the LD50 compare to the amounts found in nectar and pollen? Typically, the pollen of treated crops such as oilseed rape contains concentrations of neonics in a range from one to ten parts per billion – not much, but then these are very toxic chemicals. The amounts in nectar are usually even less, commonly below one part per billion. The big question then is: are these concentrations sufficient to harm bees? At ten parts per billion, a honeybee would need to consume about half a gram of pollen

– about five times its own body weight – to receive an LD_{50}. A bee would certainly not consume this much in a short period, although it could easily do so during its life. However, typical lab toxicity tests don't look at long-term effects; most last just a couple of days, and over this period these sorts of concentrations do not kill bees. As a result, the compounds were deemed to be safe for pollinators, and licences were granted for their use all over the world. Whenever beekeepers claimed that these compounds posed a threat to their bees, the agrochemical companies pointed to the data and argued that the amounts bees consume are not enough to kill them.

The arrival of CCD provided new impetus to investigations into bee health, and a re-examination of the hazards that bees face in the modern agricultural landscape. It led to prominent campaigns by beekeepers and environmentalists around the world, many of them targeted at getting neonics banned. In the UK the invertebrate conservation charity Buglife produced a report on neonics that argued for a ban, supported by their counterpart in the USA, the Xerces Society. As someone involved in bumblebee conservation and research, I was regularly asked to support these campaigns, but I was reluctant to do so. I wasn't aware of any compelling evidence that pinned either CCD or bumblebee decline to neonic use. It is the job of scientists, so far as is humanly possible, to be impartial and to provide the evidence that informs the decisions of others, not to become environmental lobbyists, although sometimes the distinction becomes blurred (such as when writing this book). Nonetheless this issue seemed to be one that wasn't going to go away, so in 2011 I decided to do some research of my own.

As a first step, I tried to read everything that had been published already. This was hard to do, because the work done to examine the safety of these chemicals when they were first developed is not available for scrutiny by scientists or anyone else. Various summary reports could be downloaded from European and US regulatory agencies, but these rarely gave sufficient detail to understand fully what had been done. However, there were quite a few academic studies published in the mainstream scientific literature after the chemicals became widely used, mostly carried out in the lab or with bees flying in cages. Almost all agreed that, when exposed to realistic doses of neonics such as bees might encounter on a seed-treated crop, there was little or no mortality, at least in the short term. However, some found interesting effects on bee behaviour. Neonics are, after all, neurotoxins, so it seemed plausible that sub-lethal doses might adversely affect the behaviour of bees. Studies on both honeybees and bumblebees seemed to suggest that their ability to learn and to bring food back to the nest might be impaired when fed even minute amounts of neonics. However, the effects were generally small and nothing was seen that could explain the complete collapse of honeybee colonies.

I chatted over these studies with Penelope Whitehorn and Steph O'Connor, two members of my bumblebee research group. In nature, bees travel kilometres from their nest in search of patches of flowers. They have to learn how to get pollen and nectar out of the flowers (each flower being of a different design) and then find their way home. For their nest to thrive, each worker has to do this over and over again, all day long, for days on end. Their navigational abilities are amazing. They can use the sun as a compass; they seem also able to detect the Earth's magnetic field; and they can memorise the position of various prominent landmarks such as trees and buildings. It seemed to us that what was really needed was a study of what happened to bees when exposed to neonics in a natural setting. If the exposure was impairing their mental faculties in some way, then the effects might not be at all obvious when the bees had to fly all of two metres from their nest to a dish of honey placed there for them by the experimenter.

Even a very poorly, intoxicated bee could probably manage that. On the other hand, anything that interfered with their navigation or learning would be much more likely to become a problem when faced with the challenges of the real world. If no effects were found even under such natural conditions, then I felt we might finally be able to stop worrying about neonics and look elsewhere for the cause of our bees' problems.

We sat down and devised the best experiment that we could come up with. We had no funding for this, so it had to be simple and cheap. I persuaded a colleague named Felix Wackers, of Lancaster University, to provide us with native buff-tailed bumblebee nests for free, to which he had access through his association with one of the companies that rear bumblebees for commercial use. We wanted to simulate the situation in which a wild bumblebee nest finds itself near a field of oilseed rape that has been treated as a seed with imidacloprid. Oilseed rape flowers for about one month in spring, and at this time it is a magnet for bumblebees, so during flowering one might expect a lot of the nectar and pollen coming into any nest near a rape field to be from the rape itself. Ideally we would have placed our nests next to a treated and an untreated rape field and then compared the difference, but we could not find untreated fields and we had no funds to pay for them to be planted. In any case we would have had to be certain that there were no other treated crops within flight range of a worker bee, and given that they can easily fly a kilometre or two, this was never going to be a practical option. Instead we opted to expose the bees in the lab, and then put them out in the field. That way we could control exactly what they ate.

We fed one batch of nests on clean nectar and pollen, and another on nectar and pollen carefully mixed with imidacloprid, to re-create exactly the very low concentrations found in rape.

After two weeks we took the nests out on to the campus at Stirling University and opened the doors. From that point on the bees were left to look after themselves; to gather food they would have to fly off and find flowers, just as they naturally would. We couldn't be certain that they wouldn't be exposed to more neonics in the gardens on the edge of Stirling, but no arable crops are grown nearby, so they weren't likely to be exposed to much; and at least any differences between our treatment groups would have to be due to what they fed on in the lab before they went out.

Every two weeks Penelope and Steph went out in the middle of the night (when all the bees should be at home) to weigh the nests – one dark night scaring themselves half to death by imagining that a coil of hosepipe dangling from a wooden post was in fact a very tall and sinister man in a large floppy hat.

We analysed the data as they came in, eager to see if any differences emerged between the treated and untreated nests. Slowly the average weights of nests in the two groups diverged, with the untreated nests growing at a slightly faster rate. By six weeks the differences were quite marked. After eight weeks the nests were starting to senesce, losing weight and producing males and new queens as they naturally should in summer. We collected them in and dissected them, so that we could count exactly how many eggs, larvae, pupae and adult bees there were. The results were striking. In most respects the treated nests were just a little smaller, with fewer pupae and adult bees, but in the most important respect they were dramatically different. The control nests produced, on average, about thirteen new queens. The treated nests produced an average of just two – an 85 per cent reduction. These new queens are the only stage to survive the winter, and it is the queens that found new nests the following spring. All else being equal, an 85 per cent drop in queen production means 85 per cent fewer nests being founded the following spring.

Oilseed rape is a very common crop in lowland England, so few bee nests are far from a field of it. In a recent study of the Hertfordshire landscape we found that there was almost nowhere more than one kilometre from the nearest oilseed rape, which is easily within the range of foraging bumblebees. Oilseed rape attracts honeybees and a whole range of bumblebee species, particularly buff-tails, white-tails and red-tails, plus numerous hoverflies. Almost all of it is treated with neonics – I am told that it is near-impossible to get hold of untreated seed, even if a farmer wished to. These compounds are also routinely used as seed dressings on many other crops: sunflowers, sugar beet, potatoes, wheat and maize. Well over one million hectares of UK farmland are treated with them every year. Raspberries and strawberries are sprayed with neonics during the spring and summer, using much larger amounts that are used for seed dressings, and these crops are primarily pollinated by bumble-bees. Garden insecticides are mostly based on neonics. For less than ten pounds your local garden centre will sell you a bottle of neonic containing sufficient active ingredient to kill instantly several million honeybees.* These are advertised for use on flowers, and on flowering vegetables such as beans and peas. Unlike farmers, gardeners are entirely untrained in the use of pesticides, and most probably bung on a bit extra for luck. Bees in suburban areas are playing Russian roulette every time they feed on a new patch of flowers.

Neonics are also sold as soil drenches to kill subterranean grubs that eat grass roots in lawns, golf fairways and pastures; heaven forbid that a suburban lawn should have a few brown patches where the roots of the grass have been nibbled – far better that

---

* At the time of writing a number of major garden centres and DIY chains have recently withdrawn these compounds from their shelves.

the whole lawn (and any clover or dandelion flowers it might produce) be impregnated with nerve toxins.

In urban areas trees are sometimes injected with neonics to protect them against pests; for example, avenues of trees in suburban streets may be treated to prevent outbreaks of aphids, which could result in unsightly and sticky honeydew on the cars below. An entire tree can be made toxic to all insect herbivores for several years to come by a single injection. If these trees happen to be lime trees, then bees will feed on them, adding yet more to the dose they receive.

Extrapolate our results across the country – indeed, across the world – and the likely scale of the impact on bumblebees is breathtaking and terrifying. One way or another, almost all bumblebee nests are likely to be exposed to these compounds. Perhaps we had discovered the 'smoking gun' at the root of bee declines. From being rather sceptical about the claims that neonics were wiping out bees, I found myself coming round to the view that this might well be true. We were, as you might imagine, very excited by this and keen to publish our work quickly and in a high-profile journal, where it would be noticed and acted upon. We submitted it to the journal *Science*, and waited with bated breath.

The peer-review process for scientific publications can be frustratingly slow, and it was many weeks before we got a reply. Even then it was not clear whether *Science* would eventually publish our work, for the anonymous referees to whom they had sent our paper had recommended various changes, and one of them didn't seem to think the work was particularly interesting. We did our best to comply, returned the manuscript and waited once more. Eventually, to our huge relief, *Science* declared that it would publish the work. It also revealed that it intended to publish a second, related paper alongside our own. I begged a copy and was fascinated.

It seemed that a French team, based in a government lab in Avignon, had also decided to conduct more realistic experiments on the impacts of neonics on free-flying bees. They had studied honeybees, exposing foraging workers to tiny doses of a neonic called thiamethoxam, mimicking them discovering and feeding upon a treated crop. They had glued miniature radio tags to their bees, so that the return to their hive was automatically detected and recorded by sensors mounted on the hive. Their results provided a beautifully simple explanation for the slow growth and poor performance of our treated bumblebee nests. They found that worker honeybees were much more likely to get lost on the way home if they were fed a neonic. The effects were more pronounced the further the bee was from home, and if the bee was in an unfamiliar location, from which finding its way back to the hive would require its navigation skills to be in tip-top condition. In my mind there is a simple human parallel: it is easy for a drunk to find his way home from his favourite pub, particularly if it is close to home, but put him in an unfamiliar pub and he is quite likely to get lost. The French team's findings provided the first clear indication of a mechanism that could explain the symptoms of CCD. CCD is not about bees dying; it is about them disappearing. If bees cannot find their way home, then a hive will quickly empty of bees, leaving no corpses behind. Lost bees are as good as dead. Without their hive they have no purpose in life and will quickly expire. So it seems that sub-lethal doses of neonics can indirectly kill bees in the real world, while having no measurable effect in the lab.

Science correctly anticipated that these papers, published together, would cause quite a stir. They decided to organise a press conference. The external-relations staff at the University of Stirling were quite taken aback by the idea and seemed less than enthusiastic about hosting it, so the press conference was arranged to take place in Paris instead. Press releases went out at midnight on Sunday 25 March 2012, with a strict embargo on publication of media stories until the evening of Thursday 29 – it was all terribly exciting, for someone who had not been involved in such things before. The press conference was scheduled for Thursday morning. I expected all hell to break loose on Monday, but not much happened. It took a while for the media to react, perhaps because they knew they couldn't publish anything before the Thursday night, which basically meant that the story would run in the newspapers on Friday.

After the press release went out, but before the actual papers were available to read, Defra (the Department for Environment, Food & Rural Affairs) declared that all pesticides licensed for use in the UK were perfectly safe, if used properly, and that the two new studies did not change this. This struck me as a remarkably odd position for a government department to take, particularly since it had not at the time actually been able to read the new studies. Its position seemed to reflect a belief that the systems it had in place for evaluating the safety of pesticides were infallible, and therefore that no new evidence could possibly come to light which could reveal that they had made a mistake. One might expect, and even understand, such a stance from the agrochemical companies, but from a government department – paid for by taxpayers and supposedly working on our behalf – it was mystifying.

As the week went on my phone became steadily busier, with reporters from various newspapers ringing for additional detail and quotes. Penelope and I flew to Paris on the Wednesday evening. We'd had to book a hotel way out in the suburbs as there happened to be an art festival going on in the city centre and everywhere was booked up. On the Thursday we caught the metro into the centre of Paris. The press conference was in a beautiful old building just off the Champs-Elysées, in a room that seemed more suited to a masked ball than a press conference about pesticides. Penelope and I sat nervously next to the French scientists, Mickaël Henry and Axel Decourtye, facing an audience of journalists flanked by television cameras and bright spotlights, while French aristocrats gazed imperiously down from huge gilt-framed oil paintings. We both felt like fish out of water, and were worried that we might face aggressive questions from representatives of the agro-chemical industry. Penelope was heavily pregnant at the time, and we joked that she should feign contractions if things weren't going well. In fact the conference was reasonably uneventful, although some of the questions were a little eccentric. One Mexican journalist suggested that we should develop a vaccine against neonic poisoning. Afterwards we did interviews for a couple of television stations and then escaped to a park by the Seine, where I spent the whole afternoon on my mobile fielding questions from journalists.

The next day, back in the UK, the newspapers were full of the story. The *Independent*, always a staunch supporter of bees, had put the story on the front page, while every other broadsheet and most of the tabloids had covered it in one way or another. The following day Defra's Chief Scienctific Advisor, an eminent scientist named Robert Watson, announced that Defra would review the evidence with regard to neonics. We were delighted.

In the following weeks we heard that Defra had commissioned Fera (the Food & Environment Research Agency) to repeat our study, but that they were going to expose their bumblebee nests by placing them next to real fields of rape. I couldn't see how they could find control (untreated) fields, particularly since they seemed to be starting the study immediately, but I assumed they had rather greater resources at their disposal than we had. The French Agriculture Minister announced an immediate ban of thiamethoxam (the chemical studied by the Avignon research team) on oilseed rape. The European Commission asked EFSA (the European Food Safety Authority) to conduct a thorough review of the safety of neonics. In addition EFSA launched a review of the safety tests used to evaluate new agrochemicals, particularly with a view to devising ways to detect sub-lethal effects on bees. The Environmental Audit Committee in Westminster launched an inquiry into the safety of neonics. It seemed that the science was being taken seriously by policymakers, and that wider restrictions on the use of neonics would be forthcoming.

In the meantime, research on neonics and bees continued. A paper from James Cresswell's research group at Exeter University showed that doses of imidacloprid as low as one part per billion were enough to reduce egg-laying in bumblebees by one-third. Nigel Raine's group at Royal Holloway published a study showing that one of the main effects of imidacloprid on bumblebees was that the workers collected far less pollen. These studies dovetailed neatly with our own; the slow growth and reduced queen production of our nests could readily be explained if the workers brought back less food and the queen laid fewer eggs. By the autumn of 2012 the evidence that neonics were likely to be having a major impact on wild bumblebee populations was coherent and convincing.

Sometime in October I received an anonymous email from the United States. It suggested that I should read a document that was attached. The document was entitled 'Draft Assessment Report: Initial risk assessment provided by the rapporteur Member State Germany for the existing active substance Imidacloprid, Volume 3, Annex B, February 2006'. If the title sounds dull, I can assure you that it was nothing compared to the document itself, and I was sorely tempted just to delete it, but my interest was sufficiently piqued that I eventually set about trawling through it. It comprised pages 572–790 of a much larger report, and it summarised many dozens of scientific investigations carried out by the manufacturers of imidacloprid, most of them relating to its chemical structure and the chemistry and speed of its degradation in soil. It was mind-numbingly tedious, but eventually a graph, some seventy pages in, caught my eye. It described the results of a study conducted by the pharmaceutical company Bayer on the levels of their product, imidacloprid, in soils over a six-year period in the early 1990s. They had simply sown winter wheat treated with imidacloprid for six years in a row at two sites in East Anglia, and then measured the level in the soil the day before the following application. The data were absolutely clear – the levels simply went up and up over time, up to 60 parts per billion, far higher than the concentrations we had been using with our bees. It was abundantly clear that imidacloprid has the potential to accumulate in soil if used regularly. Yet the text describing this study concluded that it demonstrated 'no potential for accumulation in soil'. What on earth was going on? I sent the document to the Environmental Audit Committee and they quickly got back to me with the opinion that it must be a fake, perhaps produced by a rabid anti-pesticide campaigner. However, further investigation showed it was genuine – and that anyone could download the original from an EC website, if they had the patience and knowledge to find it, tucked away amongst thousands of other lengthy and tedious documents.

This got me thinking. Up until this point I had been focused on bees, but had I been missing the bigger picture? I began to dig more deeply into the literature about the environmental fate of these compounds. The first thing I discovered I found quite astonishing. A study by Bayer scientists had quantified exactly how much of the active ingredient that stuck on the outside of crop seeds was taken up by the crop. The answer was: not very much – usually only about 2 per cent, and sometimes up to 20 per cent. Other studies carried out in Italy showed that about 1 per cent of the seed coating routinely blew away as toxic dust, even when the pesticide was firmly stuck to the seeds; not very much, but enough to kill immediately any honeybees flying nearby. The vast bulk of the chemicals, between 80 and 98 per cent, were ending up in the soil. The agrochemical industry had always claimed that pesticides applied as seed dressings provide much better targeting of the crop compared to those applied as sprays, but this does not appear to be true. With spray applications one can commonly get 30–50 per cent of the active ingredient on to the crop; neonic seed dressings appear to be far less efficient in this respect.

Disturbingly, studies of the persistence of neonics once in soil suggested that they could last for years. This is usually measured as a half-life – the time it takes for half of the chemical to break down – and most published estimates of the half-life put it at anywhere between 200 and 6,000 days, depending on soil type and conditions. This made sense of the East Anglian study; if quite a bit of the chemical is left in the soil after one year, then adding more every year is going to cause levels to rise over time. What does this do to invertebrates that live in soil?

To make matters worse, neonics are soluble in water – they have to be, to act systemically in plants. This would lead one to expect them to leach from soils into surrounding streams and ponds. The evidence seemed to suggest that there would be a flush of neonics washing out of soils if it rained soon after the seeds were sown (highly likely with autumn-sown crops such as winter wheat), with the remainder of the neonics binding to organic matter in the soil and then being likely to remain there for months or years. If neonics end up in clear water they are quickly broken down by sunlight, but if they settle in pond or stream sediments they can last for years. Random sampling of streams in California commonly found concentrations exceeding one part per billion – higher than the LC50 for some aquatic insects such as mayflies. Studies from the Netherlands suggest that heavy use of neonics on the bulb fields can lead to concentrations exceeding 200 parts per billion in nearby waterways, presumably sufficient to kill all insect life. I could find no similar studies from the UK of the levels in water.

So, we can expect neonics to have been gently accumulating in arable soils throughout the world for the last twenty years. It seems likely that they will be taken up by hedgerow and field-margin plants that have their roots in these soils, just as they are by the crop, meaning that they might be consumed by any farm-land herbivore – the caterpillar of a butterfly, for example. Do the nettle patches in field margins all contain neonics, so that the peacocks and small tortoiseshell caterpillars that eat them all become poisoned? Are all the moth caterpillars feeding on the hedgerow trees slowly accumulating toxins? Even if not sufficient to kill them, does this exposure interfere with their behaviour as adults, making them less able to find mates or less adept at iden-tifying the correct places to lay their eggs? Are the field margins and flower strips paid for by agri-environment schemes contami-nated with neonic dust from seed-drilling and uptake from the accumulations in soil? Neonics are also likely to be pulsing into streams in autumn when most crops are sown, and building up in aquatic sediments. What harm does this do to aquatic insects and the fish that depend on them for food? Finally, do birds and rodents eat treated crop seeds? Even though neonics are less toxic to vertebrates than to insects, a single maize seed is coated with enough neonic to kill a songbird, and seeds are commonly spilled onto the ground during sowing operations. It may be that we are poisoning the environment on a monstrous scale. Agri-environment schemes and other conservation projects are doomed to fail, if this is so. Could all of this explain why farmland wildlife is declining, despite our best efforts to look after it? I don't know, but it seems entirely plausible, and it is surely high time we found out. In the meantime it seemed to me that it might be wise to stop using neonics.

In December 2012 EFSA announced the results of its review of the safety of neonics. It seemed to agree with me; it highlighted that many of the environmental risks posed by these chemicals were not properly understood, and concluded that neonics posed an unacceptable risk when used on any crop visited by honeybees, or for crops sown when bees are active (due to the toxic dust created). A few days later the EC proposed a two-year moratorium on all such use of neonics, which was to be put to a vote of member states.

Of course the agrochemical industry didn't take this lying down. Global neonic sales are thought to be worth roughly $3.5 billion, so the people who make them have an awful lot to lose if they are banned. Our research, and that of the Avignon group, came under attack. Industry claimed that we had used unrealistically high doses, and that our work was lab-based and thus not representative of the real world. They claimed that EFSA's six-month review was shoddy and rushed. They produced glossy documents claiming that, if a ban on neonics was introduced, the EU economy would suffer to the tune of seventeen billion euros, and that 50,000 jobs would be lost, although it was unclear what evidence underlay these statistics. The document seemed to be designed to scare politicians into voting against the moratorium.

At around this time, in January 2013, I was asked to go to the Central Science Lab in York for a meeting to discuss the findings of the Fera study. This was the study that had attempted to copy our research on bumblebees and neonics, but with the exposure part of the study taking place in the field, so that the entire experiment was as realistic as possible. Fera scientists had placed buff-tailed bumblebee nests next to one of three fields, one treated with clothianidin, one with imidacloprid and a control field that was untreated with pesticides. The intention was to compare how the nests performed; if neonics are harmful to bumblebee nests, then the nests next to the treated fields ought to do worse than those next to the control field. This is not a great experimental design; one really needs several fields in each treatment, since every location will be different in numerous other ways that might confound any effects of the treatment itself. However, Fera could only find one untreated field in the whole of England, and they had had to drive 160 kilometres to get to it. As it turned out, this was the least of their problems. The bigger issue was that bees can fly, a fact of which they ought to have been aware. Buff-tails happily fly a couple of kilometres to find food, and they are suspected of avoiding foraging close to their nests, perhaps because it might attract predators. When the Fera scientists tested the food stores in the nests for pesticides, it turned out that the 'control' nests contained a selection of neonics. The bees had clearly found a crop treated with thiamethoxam, and another treated with clothianidin (or perhaps wild flowers contaminated with both) and had been feeding on these. The control nests were exposed to just as much pesticide as the 'treated' bees. This was a disaster, and it might have been wise to abandon the experiment at this point and start again. All that could really be learned was that bumblebee nests in UK farmland are heavily exposed to a cocktail of chemicals, no matter where you put them. Instead, Fera or their Defra bosses chose to 'publish' the study by placing it online, and declared that it demonstrated there was no major effect of pesticides on bumblebees. The normal scientific process is to submit studies to a peer-reviewed journal, where they are scrutinised by anonymous, independent experts. Weak and flawed studies such as this are weeded out, providing a degree of quality control. Simply placing results on the Internet is not the way science proceeds.

In fact, if one looks closely at the Fera study, it emerges that there is a clear and strong correlation between the levels of pesticide in each nest and how poorly the nest fared, in terms of growth and queen production. Nests with more pesticide grew more slowly and produced fewer queens. Oddly this was not mentioned in the summary of the paper, or when Defra subsequently referred to the study.

The EC proposal went to a vote on 15 March 2013; of twenty-seven EU states, thirteen voted for the moratorium, nine against and five (including the UK) abstained. The proposal was one vote short of an absolute majority, and so by EC rules the proposal was rejected. Defra stated that there were still too many unknowns for it to formulate an opinion. With impressive chutzpah, the government's Chief Scientific Advisor, Sir Mark Walport, stated that everyone else had misinterpreted the evidence, and that in the meantime we should invoke the 'precautionary principle' by continuing to use neonics. It cited Fera's study as showing that there seemed to be no major effect of neonics on bumblebees, even though it was well aware that Fera's study was a disaster.

It was hard for those of us involved to understand the stance of countries that voted against the proposal, or those that abstained. Almost everybody agrees that there are huge gaps in our understanding of the risks associated with these chemicals, and that they were not adequately assessed when first proposed for use. EFSA's team of scientists concluded that these chemicals pose unacceptable, but poorly quantified, risks to bees. If neonics were brought to the market for the first time now, supported by the statement 'These chemicals seem to pose substantial risks to the environment; we haven't properly evaluated these, but we'd like to start selling them all over the world anyway', they would of course be rejected. But because they slipped through the net and are already in use, there seems to be a reluctance to admit a mistake, or to upset the status quo. Somehow the UK government regarded its own stance as an adoption of the precautionary principle, when everyone else's understanding was that adopting the precautionary principle would lead us to do the opposite – i.e. stop using toxic chemicals until we could be sure they were safe, rather than vice versa.

Fortunately the EU Health and Consumer Commissioner, Tonio Borg, decided to bring the proposal back for a second vote on 29 April. Once again the UK voted against the ban, but this time Germany switched to a vote in favour, and the bill was passed by fifteen votes to eight. Whatever one's views on EU membership, this seems to me to be a situation in which common sense was imposed upon an unwilling UK by our more sensible continental neighbours.

One question that is often raised is what will farmers use to control the pests of crops such as oilseed rape, now that neonics are going to be withdrawn (at least temporarily)? It seems to me that we should return to IPM, minimising the use of pesticides by monitoring pest problems and boosting the numbers of natural enemies, using chemical controls only when they have to be used. Prophylactic use of persistent pesticides is not a sustainable approach because it leads to pesticides accumulating, damages the populations of bees and the natural enemies of pests, and is highly likely to lead to the evolution of resistance in the pests, as has already happened in the USA, where some Colorado-beetle populations are nearly immune to neonics. Exactly the same argument explains why doctors are reluctant to prescribe antibiotics, and why they would never prescribe them prophylactically to avoid illness – if they did, bacteria would swiftly evolve resistance and the drugs would no longer work.

I would also question whether farmers always need to replace neonics with something else. The evidence that crop yields directly benefit from using neonics is surprisingly hard to pin down. Oilseed-rape yields are no higher now than they were in 1991, when no neonics were available. Recent studies of soya-bean farming in Brazil suggest that farmers would obtain the same or greater yields for less expenditure on pesticides if they switched from using neonics to using IPM. Studies from the United States have shown that soya-bean yields are not improved one jot by using neonicotoids. Yet in both Brazil and the USA nearly all soya-bean farmers use neonics.

Why would farmers use chemicals if they don't need them? Much of the advice given to farmers on what chemicals to use comes from agronomists, most of whom work for large companies involved in supplying pesticides. This is hardly independent advice, and is surely prone to overselling. If a farmer is advised that product X will provide his crop with the best protection, he is likely to buy it; it would be a brave person who ignored the advice of his agronomist. Yet he has no way of knowing that the agronomist isn't recommending product X because it gives him the best profit. Farmers growing oilseed rape are generally recommended to use a neonic-dressed seed, but they are also told to spray several times in the autumn, spring and summer with other pesticides such as pyrethroids, so what function is the seed dressing serving?

There are some clear lessons to be learned from all of this. Safety tests for new generations of agrochemicals need to be realistic, taking account of subtle sub-lethal effects that cannot be revealed by lab tests. As the French *Science* paper shows, even putting a hive immediately next to a treated flowering crop will not reveal much in the way of impacts if the compounds affect the bees' ability to navigate over long distances. And safety tests are currently carried out by the agrochemical companies themselves, or by private companies whose main income derives from them. Does anyone really think that a company that has invested millions in developing a new chemical will be entirely unbiased when presenting its safety-testing data to the authorities? As someone who carries out experiments and analyses data all the time, I am aware that it is all too easy to influence the outcome of experiments, either accidentally or deliberately, and that data can always be analysed in a number of different ways, and these may not all give the same answer. And of course if the 'wrong' result emerges, it is easy to argue that the experiment was flawed in some way and that it should be repeated. I have no evidence that such things go on, but it would be surprising if they didn't, with so much money at stake. All such tests and analyses should surely be carried out by strictly independent bodies.

It is just over fifty years since Rachel Carson wrote *Silent Spring*, in which she highlighted how agrochemicals were devastating wildlife. At the time DDT and its organochlorine relatives, dieldrin and aldrin, were the main culprits. These chemicals have long since been banned in most developing countries in favour of safer chemicals. Are neonics so different? The influence of big industry seems to have pushed farming far away from the sustainable approaches that I was taught about in the 1980s.

At this point in time it is hard to predict what will happen in the EU after the proposed two-year ban expires. Neonics will continue to be used extensively for non-flowering crops such as winter wheat, and with oilseed rape generally being grown in rotation with wheat, there are still likely to be neonic residues in the nectar and pollen. Even if we completely stopped using neonics, they would be in our soils for years to come. So any benefits from the partial moratorium will not be apparent within two years. In any case there seems to be no plan to monitor the benefits, so if they did occur (which is unlikely) we wouldn't even know. All in all, it seems an ineffectual half-measure, a political manoeuvre perhaps intended to keep as many people happy as possible, but not one that has resolved the situation satisfactorily or permanently.

I think environmentalists will triumph on this issue in the end. But new generations of neonics are rumoured to be in development, to which the current moratorium does not apply, and no doubt other compounds too. Environmentalists may win this battle, but they are losing the war, and wildlife is steadily paying the price. Until we fundamentally change the system for testing and approving new chemicals, and for advising farmers, how long will it be before something worse comes along? In the past we were assured that organochlorides were safe, until it transpired that they weren't. Then we were told that organophosphates were all safe, until many of them were banned. Now Defra repeats the mantra that all pesticides are subject to rigorous safety testing, and that all of the ones that are currently licensed are perfectly safe. History suggests it is wrong. It is remarkable and sad that, half a century on from *Silent Spring*, we don't seem to have learned any lessons at all.

I must add one final note of caution. I am not claiming for one second that neonics are the only problem that bees or other wildlife face in the modern world. Bee declines are undoubtedly due to a mixture of factors, probably including diseases, *Varroa* mites (in the case of honeybees), lack of flowers, a monotonous diet and exposure to multiple pesticides, all providing a potent cocktail of stressors. It is very likely that these factors interact; bees that are mildly poisoned will be more susceptible to disease, less able to cope with starvation, and so on. We can't easily solve all of the problems affecting bees, but we can stop poisoning them. Isn't it time we did so?

CHAPTER FOURTEEN

The Inbred Isles

8 July 2013. Run: 40 mins 48 secs. Slow again today, perhaps a little too much Côtes du Rhône last night. People: none. Dogs: 7. Butterfly species: 8. Damp morning; there was heavy rain in the night and ominous, bruised clouds to the west suggest more to come. I nearly trod on a young fire salamander that was taking advantage of the humidity and trundling slowly across the track near the Épenède water tower, a wonderful creature with a glossy black skin marked with golden-yellow streaks and spots. I placed it in the long grass for safety – perhaps I should knock ten seconds off my time to allow for salamander rescue?

In the 1960s, some of the oddest ecological experiments ever to be performed were carried out by Daniel Simberloff in the Florida Keys. Dan was a grad student under the legendary E.O. Wilson, one of the great champions of the importance of insects in natural ecosystems and one of my personal heroes. Dan's doctoral studies focused on testing some of the predictions of 'Island Biogeography Theory' (known as IBT to its close friends). IBT was an idea hatched by E.O. Wilson and his collaborator Robert MacArthur to try to explain the relatively small numbers of species found on oceanic islands compared to the mainland. MacArthur and Wilson's theory is one of those things that now seem fairly obvious, but at the time it was proposed it was seen as ground-breaking and revolutionary.* It had long been known that small, remote islands tend to have few species, whereas big islands and those very close to the mainland tend to have more. MacArthur and Wilson argued that the number of species on any island was a balance between immigration and extinction. Animals and plants occasionally get blown or washed off the mainland, and are more likely to arrive in one piece at islands close to where they came from, or on big islands (simply because they are more likely to bump into them). Thus one might imagine that a big island just offshore from a major continent might be exposed to a constant stream of potential colonists washing up (immigration is high). In contrast, a tiny island in the middle of the Pacific is likely to receive few colonists (immigration is low). In addition, big islands are likely to have a greater diversity of habitats, such as mountains, forests and lakes, while small islands will have few. Thus colonisers arriving on a big island are more likely to find somewhere suitable to live.

MacArthur and Wilson also proposed that the arrival of new species on islands will be offset by extinctions of existing species, so that once an island has been in existence for a while, we would

---

* I hesitate to use the phrase 'paradigm-shifting', which sprang to mind here but is now horribly overused in scientific circles – it seems that every research grant application has to pretend that it is going to shift at least half a dozen paradigms if it is going to stand any chance of receiving funding. The European Research Council actually specifies that it will only fund paradigm-shifting research, but as far as I can see, if you know you are going to shift a paradigm before you have done the work, it must be a pretty dumb paradigm in the first place. When I was an undergraduate I embarrassed myself considerably by pronouncing paradigm as para-dig-m (in case, like me, you don't know, it should be para-dime).

expect there to be an approximately stable total number of species, with occasional extinctions balancing occasional new arrivals. Extinctions of island populations naturally occur every now and again, due to a multitude of causes – disease, storms, and just plain bad luck. Extinction is more likely, the smaller the population; in a big population there is more chance that a few lucky individuals will survive whatever catastrophe has struck. This pattern is likely to be exacerbated by inbreeding; in small populations it takes just a few generations for everybody to become related to everybody else. For reasons to which I will return, it is not a good idea to marry your brother or cousin, but you may have no choice if you live on a small island. Inbreeding is likely to weaken a population, again making small islands prone to high extinction rates.

You may wonder why I am telling you all this, but the relevance will eventually emerge, so bear with me. To recap, few new species ever arrive at small, remote islands, and when they do they are quite likely to go extinct there before too long, and this is essentially why these small, remote islands tend to have fewer species than large islands near a mainland source of colonists. Now this all seems like an eminently logical theory, but it is hard to test. You would need to manipulate the size or remoteness of an island, which of course isn't possible. Unless you are Daniel Simberloff. Dan hit upon the extreme idea of manipulating the size of mangrove islands off the coast of Florida. Mangroves are salt-tolerant trees that grow in the shallow, muddy waters off tropical coasts, and they support a wealth of insect, crustacean and bird life, as well as providing a spawning ground for numerous tropical fish. Clumps of trees grow wherever the sea is shallow and sufficiently sheltered, so the Florida coast is dotted with mangrove islands of various sizes and shapes, and of varying distance from the shore. Dan couldn't easily make these islands bigger or move them further offshore, but with the enthusiastic deployment of a chainsaw he could certainly make them smaller, and this he proceeded to do. He found, as you will by now expect, that if he shrank the size of an island, he increased the rate of extinction, so that the number of species quickly dropped and then reached an equilibrium. This supported part of MacArthur and Wilson's theory – the bit to do with island size and extinction rates – but left untested the parts relating to colonisation rate. How could one study the rate at which new species colonised an island?

Dan devised another boldly destructive approach to tackle this question. He enlisted the help of a local professional exterminator named Steve Tendrich, who one might imagine must have thought this a highly eccentric project. Together they constructed scaffold frames over entire mangrove islands (admittedly rather small ones – some were just a few metres across). Plastic sheeting not being available at the time, they hung rubberised netting over the scaffold to encase the islands and then pumped in highly toxic methyl-bromide gas, fumigating the little islands and killing everything but the mangroves themselves. They managed to repeat this on six small islands at varying distances from the mainland (not a great sample size, but quite impressive, given the effort involved). After denuding the islands of their fauna, Dan monitored how quickly they were recolonised over the next two years. Lo and behold, islands near the mainland were quickly repopulated with a range of species, while those further offshore were colonised more slowly and never reached the level of species richness that was rapidly recovered on the inshore islands.

If MacArthur and Wilson's theory only applied to oceanic islands, it would never have created the stir that it did, and I would not be writing about it now. However, it quickly became apparent that the theory could be applied to habitat 'islands' separated by a 'sea' of unsuitable habitat. Obvious natural examples might include ponds or mountaintops, each of which supports specialist creatures that cannot survive for long in the intervening habitat. More importantly, man's activities have created habitat patches – fragments of natural or semi-natural habitat surrounded by a sea of heavily modified land, such as farmland or urban areas. The large tracts of woodland that once covered much of Europe in a near-continuous blanket now exist as fragments, separated by fields of crops that are inhospitable to most woodland inhabitants. Flower-rich grasslands such as the meadow at Chez Nauche were once scattered thickly across the countryside so that no patch was far from any other patch, but they are now few and far between.

Of course habitat islands are not quite the same as oceanic islands; for most animals, crossing an arable field is much easier and less daunting than crossing an expanse of sea, so – all else being equal – we might expect habitat islands to have more colonisers than oceanic islands. Nonetheless the broad principles remain and have profound consequences for conservation, for they make predictions as to how many species a nature reserve might be likely to support, and how we might prioritise conservation efforts and expenditure.

Conservation efforts are limited by funds, and are constrained by other pressures on the land. In our modern, crowded world we can only have so much in the way of woodlands and meadows, for we need to grow food and build houses, out-of-town supermarkets, industrial estates and roads.* Supposing we are faced

---

* Actually, no we don't – at least not the roads part. It is a source of constant dismay to me that successive governments unthinkingly accept forecasts of future growth in traffic, and hence the need for endless

with a situation in which we can only save, say, 1,000 hectares of woodland, would we be best to save one big wood or dozens of little ones? In 1975 Jared Diamond used Island Biogeography Theory to argue that the former would be the best strategy, for the larger patch would support more species. I'm sure he didn't realise it at the time, but in doing so Diamond helped to spark one of the longest-running debates in ecological history, one that has rumbled on in various forms till today. It became known as the SLOSS debate: 'Single Large Or Several Small'.

Perhaps surprisingly, given that Diamond was essentially building on Dan Simberloff's work, Simberloff himself was one of the first to disagree with Diamond, pointing out that the logic only worked if small habitat patches simply contained subsets of the species found in a larger habitat patch. In reality, this is rarely likely to be the case. Suppose, for example, that we had the job of prioritising woodland conservation in the UK – all but 1,000 hectares was to be swept away to provide extra parking. Which bit(s) should we save? One huge chunk of, say, the New Forest, or lots of little patches, from the twisted, lichen-encrusted oaks of Wistman's Wood on Dartmoor, to a bit of ancient Caledonian pine forest at Abernethy in the Cairngorms and a patch of Breckland forest on the sandy soils of Norfolk? In this example, one would certainly save many more species by having lots of little patches, for these woods all have different characters and each supports a different range of species.

---

road-widening schemes, bypasses, and so on. Why aren't we spending this same money encouraging people out of their cars and on to public transport, or giving incentives to companies to allow their staff to work from home one or two days a week? This would also reduce pollution, and would reduce the need to grow biofuel crops, freeing up more land for food production or for conservation.

Maybe this example is a little unfair, and it is certainly not a very plausible scenario. So let's suppose instead that we had a single chunk of reasonably homogeneous habitat – let us say one large flower-rich meadow – and that much of it was to be lost to make way for a housing development. Would we be best to save one large piece, or lots of little ones scattered among the houses? This seems like the sort of scenario that planners might regularly face in the real world. David Quammen writes beautifully about the SLOSS debate in the *The Song of the Dodo*, in which he coins the Persian Rug analogy. Imagine we take a ten-foot-by-ten-foot Persian rug and cut it into 100 pieces. Do we get 100 perfect replicas of the original? Of course we do not; instead, we get 100 worthless, frayed scraps of carpet. The same argument could be applied to the meadow: the little patches would be trampled by the residents, invaded by garden weeds and would be unlikely to support much more than a handful of the original diversity of life in the meadow.

These two arguments lead to opposite conclusions: the first that we should save lots of small fragments; the second that we should save large, intact tracts. Resolving this difference occupied quite a number of the world's ecologists for several decades in the later twentieth century, and there is still no clear agreement, although interest in the debate has finally waned.

Of course to some extent all of this misses the point. In the real world we save what we can where we can, and generally make do with what is possible, rather than worrying about what would be ideal. The housing planners are likely to be much more driven by the practicalities of road access, drainage and so on, when deciding where to put the houses, than they are with worrying about biodiversity. Where a habitat has been largely lost we might try to re-create patches of it, but the size and location of such patches are rarely determined by any ecological theory, but rather by what is available. When I bought Chez Nauche I did not have unlimited choice as to where it would be, or how large the meadow. As I had very limited funds, its location was largely determined by the cheapness of property and land in the Charente compared to other parts of France, and by what was for sale. Nonetheless, many of the ideas that emerged from the arguments over Island Biogeography Theory are useful. For example, the distance from a source of colonists plays a role in deciding how to manage my meadow. If there were a lovely flower-rich grassland in one of the neighbouring fields, there would be a ready source of wild-flower seeds and the recovery of my meadow to a flower-rich state would have been rapid, just as Simberloff's fumigated islands were colonised quickly when they were close to shore. Unfortunately there was not, so I have had to collect seeds from some flowers – particularly those with heavy seeds and no long-distance mechanism for dispersal – and sprinkle them in the meadow (avoiding, of course, my experimental plots, which would mess up my experiments).

There is another way in which we can increase colonisation of nature reserves and so boost the number of species they are likely to have, and this is by providing habitat corridors along which animals can move. I have already mentioned that many animals are likely to be unwilling to cross water, but may be much more likely to cross land, even if it is inhospitable. However, this varies enormously between species. Some birds and butterflies will happily zoom across cereal fields, towns or busy roads to get from one meadow or woodland to another, but many will not. For example, some woodland bird species, such as long-tailed tits and dunnocks, are unwilling to leave their preferred habitat and fly out into the open, and it is these species that are likely to be most badly affected by habitat fragmentation. We can make life much easier for such creatures if we provide links between patches of habitats; in the case of woodland, hedges often provide routes for dispersal for creatures that shun open fields. Tunnels under roads have been used with some success to help creatures as diverse as elephants and badgers to get from one habitat patch to another.

Other creatures are harder to accommodate. For instance, some butterflies found in flower-rich meadows are enormously sedentary – particularly blue butterflies such as the Adonis and small blue, both of which will often happily live out their entire life in a few square metres of downland, and will rarely (if ever) choose to head off into the unknown on a perilous adventure. One can hardly blame them these days; their chances of finding another suitable patch of habitat are remote, so they are probably wise to stay put. On the other hand, their sedentary lifestyle cuts them off from gene flow – each small population in a habitat fragment becomes entirely cut off from all the others, and this can doom it to eventual extinction.

The problem with isolation is that it leads to inbreeding. Small populations tend to have less genetic variation than large ones to start with, simply because there are fewer copies of each gene. They also tend to lose genetic diversity over time, through a process known as 'genetic drift', whereby rare forms of genes drop out of the population by chance. With less genetic diversity, the population is less able to adapt to any change in the environment. What is more, inbreeding can lead to the expression of rare, harmful, recessive genes. Every individual has a few dud genes; humans have somewhere in the region of 50,000 genes (though probably far fewer that are absolutely vital to our well-being), of which on average about four are likely to be faulty, but because we have two copies of every gene, and as long as at least one copy is okay, we are fine. Fortunately your partner is unlikely to have the same faulty genes that you do, so there is little chance of your offspring getting two non-functioning copies of the same gene, although it does occasionally happen. If they do, they may be severely ill, disabled or even die early in development. However, if you are related to your partner, then you are likely to have some of the same faulty genes. This is not good, for then, for every faulty gene you have in common, there is a one-in-four chance that your offspring will inherit both non-functioning copies. As I have mentioned, in a small population it takes only a few generations before everybody becomes cousins of one another, sharing grandparents or great-grandparents. In lab studies using insects, forced inbreeding over a few generations quickly results in a population of feeble individuals with low fertility and reduced life expectancy, a phenomenon known as 'inbreeding depression'. Even without this, small populations are much more likely to go extinct than large ones just through bad luck, but when these small populations are also subjected to inbreeding they are unlikely to persist for long.

Although the effects of inbreeding are well established from lab studies, until fairly recently there were few good examples of inbreeding hampering the survival of insect populations in the wild. The first – and still one of the most elegant – demonstration of the negative effects of inbreeding on insects was provided by Finnish ecologist Ilkka Hanski's work on populations of the Glanville fritillary, one of the most common butterfly species in my meadow. In Finland, Glanville fritillaries are found only in the Åland archipelago, a cluster of more than 6,000 small islands in the Baltic Sea, halfway between mainland Finland and Sweden. On these islands the Glanville fritillary is widespread and fairly common, occupying dry, grassy meadow areas where its food plant, ribwort plantain, is to be found. The lovely adult butterflies with their orange-and-black chequerboard wings emerge in May and June and lay batches of eggs which, when they hatch into small black spiny caterpillars, live in gregarious clusters within protective webs that they spin among the plantains.

Hanski's team (which consists largely of undergraduate students) has repeatedly surveyed about 4,000 of these meadows every year since 1991, recording whether or not the butterfly is present in each patch – an astonishing effort, but presumably involving many happy hours of splashing about in boats to get between the mostly uninhabited islands. Their work has become a classic study in what is known as 'metapopulation dynamics', and their findings have a lovely symmetry with MacArthur and Wilson's theory.

Hanski's group has been studying just one species while MacArthur and Wilson were trying to understand what determined the number of species in entire communities, but the processes involved are exactly the same. Hanski found that the majority of the 4,000 meadows had no Glanville fritillaries in any particular year, with only 300–500 occupied ones. Some of these populations are tiny – just a few butterflies. Populations in any particular patch go extinct quite frequently, but these extinctions are roughly balanced by colonisation events whereby butterflies discovered and established a new population in an empty patch. A large regional population composed of lots of small fragmented populations, each of which blink in and out of existence from time to time, is what is known as a 'metapopulation'.

Just as small and remote islands have fewer species in them in part because they are less likely to be bumped into by wandering creatures, so Hanski's habitat patches were less likely to be colonised by butterflies (and so tended to remain unoccupied for longer) if they were small, or a long way from other patches. Similarly, he found that small populations were more likely to go extinct. Extinctions were caused by a variety of events – outbreaks of parasitoid wasps, overgrazing and trampling by cattle, human disturbance, and so on, but all of these were more likely to prove fatal to a local population if it was small. What is more, populations were found to be more likely to go extinct if they lacked genetic diversity – the first time that inbreeding had been shown to be harmful to wild populations of any insect. Populations with low genetic diversity were found to have reduced survival of both caterpillars and adults, and a low hatching rate of eggs, factors that are obviously likely to push a small population over the edge to extinction.

This finding was followed up by one of Hanski's students named Marko Nieminen, who created new wild populations from either the offspring of unrelated individuals or a mating between a brother and sister. The latter proved much more likely to die out, mainly because the low egg-hatching rate led to smaller groups of the gregarious caterpillars, and smaller groups are known to have a much-reduced chance of surviving through the winter.

There are reasons to believe that social insects such as bumblebees might be particularly prone to inbreeding, and this might partly explain why some bumblebee species have fared so poorly in recent years. In butterflies such as the Glanville fritillary – indeed, in most creatures – every adult individual tries to breed, and although quite a few fail, many pass on their genes to the next generation. In bumblebees this is far from the case. The large majority of the population are workers, most of which will not produce any offspring. Save for a few sneaky workers that try (largely unsuccessfully) to produce sons, reproduction is the preserve of the queen. This means that the breeding population – the gene pool – is much smaller than you might at first think. Each nest has just one breeding individual (who also carries inside her the sperm of her mate). So ten nests, which might collectively contain several thousand bumblebees at their peak, actually represent just twenty breeding individuals. A viable population, one that will be big enough that it does not suffer from inbreeding, needs to contain several hundred breeding individuals – let's say at least 100 nests. We don't know for sure how big an area of habitat would be needed to support 100 bumblebee nests of any particular species – it would depend on how good the habitat was, and would probably vary between bumblebee species – but it is likely to be in the region of hundreds of hectares. In the UK, indeed in most of western Europe, most nature reserves are smaller than this. For example, most of the patches of flower-rich grassland that survive consist of single meadows, often of no more than a dozen hectares or so, surrounded by arable crops. In other words, the habitat islands that remain are likely to be too small to support a viable bumblebee population. Lots of such small islands, close together, could support a metapopulation just like that of the Glanville fritillaries in the Åland archipelago. But we don't have lots of patches of flower-rich grassland, so it is likely that fragmented populations of our rarer bees are suffering from inbreeding.

In bumblebees, inbreeding can lead to some rather peculiar gender-bending side-effects. To understand them requires a little diversion into how sex is determined in bumblebees. In many animals (including ourselves) sex is determined by which sex chromosomes we inherit. If we get an X and a Y, we are male; two Xs and we are female (some animals do it the other way round). In bees, it is quite different; sex is determined by a single gene. If an individual has two different copies of this gene, it is female. If it has two identical copies, or just one copy, it is male. Female bees, like us, have two copies of each chromosome – to use the technical term, they are diploid. In a genetically healthy population there are usually lots of different versions of the sex-determining gene, so the chances are that diploid individuals will have two different copies and thus will be female. Male bees, typically, have just one copy of each chromosome – they are haploid. To produce a son, a female bee has just to lay an unfertilised egg; the haploid gamete develops into a healthy son. This means that male bees have no father. To produce a daughter, the female bee fertilises her egg using sperm from a male; in bumblebees this sperm had been stored inside the queen since she mated the previous summer. So long as the copy of the sex-determining gene in the sperm is different from each of the two different copies held by the mother, then these diploid offspring will all be female.

The problem arises when the population is small. As I have said, small populations tend to lose genetic diversity through drift, so the number of versions of the sex-determining gene declines over time. Also, individuals soon become related to one another, so the chances that one of the queen's sex-determining genes and the male's match becomes more and more likely. If they do, then half of the queen's diploid offspring will have two identical sex-determining genes. This is a disaster for the nest, for these individuals develop as 'diploid males', which do no work for the colony. Essentially half of the colony's workforce is useless – worse than useless, since they place a burden on resources, eating the food stores and not replacing them. This might not be so bad if the diploid males went out and mated with new queens, but diploid males seem to have lower fertility than normal males, and in any case the nest starts to produce them right at the beginning of the season, when there are no virgin queens around to mate with. With half their workforce sitting around idly scoffing food, we might expect such nests to die out long before other nests start to produce new queens in high summer.

Oddly enough, this is not how it seems to work out, and we don't yet know why. The presence of diploid males can be used as a warning sign of inbreeding in bees, and my PhD student Ben Darvill screened populations of some rare species, such as the moss carder bee, to see which populations are becoming inbred. He focused on the Hebrides, another island system where agriculture has changed relatively little and patches of flower-rich grassland are common. On the smaller, more remote islands of the Hebrides such as the Monach Isles, a tiny cluster of beautiful, sandy-shored islands off the west coast of North Uist, he found that diploid males were quite common. The unexpected thing was that some of the remaining bees were triploids – they had three sets of chromosomes.

Triploids look and act like female bees. They are presumably produced by diploid males successfully mating with a normal, diploid queen, against all the odds. This requires a nest producing diploid males to survive long enough to produce these males in late summer when new queens are on the wing, and then for the diploid males to outcompete normal, healthy haploid males in finding and courting a virgin female, and finally to produce viable (but diploid) sperm, despite their supposed low fertility. On the Monachs one in five of the female bees turned out to be triploids. Clearly diploid males are not quite as hopeless as was thought. Nonetheless, we now have other evidence that the inbreeding that leads to diploid male production also has other harmful side-effects. Small, isolated bumblebee populations have measurably lower genetic diversity than the larger populations, and rare species tend to have much lower genetic diversity than the common ones. What is more, Penelope Whitehorn has found that the more inbred populations suffer from higher loads of the internal parasite *Crithidia bombi*, presumably a sign of their generally low vigour. We do not yet have a data set to compare with Hanski's butterfly study, but I would hazard a guess that these small, isolated bumblebee populations, with little genetic diversity, high levels of diploid males and high parasite load, are probably not long for this world. However, if the population on the Monach Isles disappears it would not be a disaster as there are still plenty of bigger, less-isolated habitat patches on the larger Hebridean islands. The Monach Isles might well be recolonised in the future, and we would predict that the moss carder bee population there will continue to blink in and out of existence every few years, just as many of Hanski's butterfly populations do.

There are important lessons to be learned from all of this. The survival of the metapopulation as a whole depends upon colonisation events balancing extinctions. If the extinction of populations occurs faster than new ones are created, then the number of occupied patches will slowly decline over time, until eventually there are none. In the Åland islands, where the human population is low and agriculture is not intensive, there are still lots of habitat patches, and many of them are close together. Hence it is easy for unoccupied patches to be recolonised, and so new populations spring up as fast as old ones become extinct. Overall, Glanville fritillaries are doing just fine in the region. Similarly, moss carder bees are doing okay in the Hebrides. But imagine what would happen if a few of the habitat patches are destroyed. The average distance between patches of meadow will then be a little greater, slightly reducing the colonisation rate. If there are fewer meadows in total, there will be fewer occupied patches, and hence fewer sources of colonists. Hence the proportion of the remaining meadows that have butterflies or bumblebees at any one time will be lower. Remove a few more meadows, and it will become lower still. The remaining populations will be few and far between and will begin to become inbred, because they will receive few visitors from outside, and hence extinction rates will begin to climb.

Remove too many meadows and you will reach a tipping point, beyond which the number of occupied patches will gently but steadily decline to zero. This is known by biologists as 'metapopulation collapse'.

Unfortunately it is probably a very common process. Most natural habitats in Europe are now highly fragmented, and the fragments are often not close together. Hence many populations of sedentary habitat specialists – be they snails, butterflies or bumblebees – are separated from one another, with levels of inbreeding slowly increasing. Many might be doomed to extinction by what we have done in the past, so that even if no further habitats are lost we can expect them to slowly, inexorably disappear. This is sometimes described as an 'extinction debt', and in truth we do not know how many species in the world are in this situation.

Just as it is true that no man is an island, so no habitat island is truly isolated. It cannot survive on its own, for it depends on other islands for sources of colonists, for gene flow to keep populations healthy. Chez Nauche is far more diverse that it was when it was a cereal field, but how many more species might have arrived 100 years ago, when there were many flowery meadows nearby? There are no large blue butterflies, or corncrakes, because these species have largely gone from the landscape – there aren't enough patches left for them to survive. No matter how carefully a nature reserve or habitat patch is managed, it is really only as good as the network of patches of which it is a part.

What we do know is that it is possible to help. We can create new habitat islands, filling in some of the gaps, making the network of patches a little more connected. We can plant new woodlands, link them with new hedgerows, and regenerate flower-rich grasslands such as the meadow at Chez Nauche. It takes time, and we need lots of them, so all the more reason to start now. There is an old Chinese proverb: 'The best time to plant a tree was twenty years ago. The second-best time is now . . .'

CHAPTER FIFTEEN

Easter Island

1 August 2013. Run: 39 mins 38 secs. I was up early this morning; the mist hadn't cleared as I set off and as I ran up the drive one of the barn owls swooped silently past, neither of us seeing each other until we almost collided – a wonderful, ghostly creature. People: only Monsieur Fontaneau junior, inspecting his cows. Dogs: 6. Butterfly species: 8, including a large tortoiseshell nectaring on a spear thistle. This butterfly is sadly extinct in the UK, wiped out by the demise of the elms – its larval food plant – at the hands of Dutch elm disease.

> *I was taught that the human brain was the crowning glory of evolution so far, but I think it's a very poor scheme for survival.*
>
> Kurt Vonnegut

Ninety thousand years ago a group of humans living in Africa decided to go for a walk. It was a long and slow walk, fraught with danger. It would take them and their descendants about 80,000 years, but it was perhaps the most significant trek in the Earth's history. For obvious reasons we don't know many of the details – they have been pieced together by archaeologists from fragments of bone and shards of rough stone tools excavated from thousands of sites around the globe. There will no doubt be significant new discoveries and endless arguments about the specifics, but what follows is probably somewhere near correct.

Modern humans – apes belonging to the species *Homo sapiens* – evolved in Africa perhaps 160,000 years ago, at a time when much of Europe, North America and Asia was locked under vast sheets of ice. Our ancestors remained more or less confined to Africa for 70,000 years and then, for reasons at which we can only guess, a group of them left Africa, crossing the mouth of the Red Sea on to the Arabian Peninsula, carrying their stone tools: axes, knives, hammers and arrowheads. They seem to have had a close association with the sea, for they stuck to the coast and spread slowly eastwards, successively occupying the coasts and islands of India, Thailand, Malaysia and Indonesia. It took about 30,000 years before some of them crossed to New Guinea and Australia, approximately 60,000 years ago. Fifty thousand years ago the climate began to warm, and humans were able to spread northwards into the Middle East and Europe. The first humans colonised Britain perhaps 40,000 years ago. At the same time we were also spreading north-east, into Central Asia and China. Twenty-five thousand years ago we had occupied most of the old world, from Britain and Spain in the west to Tasmania and the Bering Straits at the far eastern tip of Siberia. Shortly afterwards a particularly hardy group of humans crossed the eighty kilometres or so from Russia to Alaska, and so colonised the Americas. The climate entered another cold period (the last 'glacial maximum', in geologists' terms) and buried much of northern Europe, Russia and North America under a vast sheet of ice, pushing us southwards for a while and chasing the new arrivals in America down into the southern states, Central America and beyond. We arrived in Chile perhaps 12,000 years ago, so that about 80,000 years after our ancestors first set out from Africa, all of the Earth's major land masses had been occupied by humans, with the exceptions of inhospitable Greenland and Antarctica.

The final wave of colonisation, of the most remote but habitable places on Earth, took place rather later – perhaps 1,000 years ago, when adventurous Polynesians in dugout canoes chose to sail east from Asia into the unknown vastness of the Pacific. They discovered and colonised New Zealand, the various small archipelagos of Fiji, Samoa and so on, and eventually made it to Hawaii and finally to Easter Island, arguably the most remote inhabited island on Earth.

Homo sapiens were not the first hominids to leave Africa. Other hominids such as *Homo erectus* and *Homo neanderthalensis* (Neanderthals) had got a huge head-start, spreading throughout much of the old world long before we arrived – *Homo erectus* was found all over Europe and Asia well over one million years ago, while Neanderthals spread throughout Europe perhaps 600,000 years ago. At least twelve different species of ape belonging to the genus *Homo* have so far been described, and it seems very likely that there are many more awaiting discovery. The tiny *Homo floresiensis* from Indonesia, weighing in at about twenty-five kilos and standing just one metre tall, was not discovered until 2003.

The world that early hominids colonised was dominated by large mammals. Following the extinction of the dinosaurs some sixty-five million years ago, widely thought to have been the result of the devastation caused by an asteroid striking the Yucatan peninsula, the few small mammals that survived went on to proliferate. Many new species arose and filled the vacant niches once occupied by dinosaurs, some of them becoming huge. In the Americas this 'megafauna' included giant sloths, camels and llamas, many species of bison, moose and ox, giant beavers, mammoths and mastodons. These herbivores were predated by some of the most formidable predators to walk the Earth in the last sixty-five million years, including huge two-tonne short-faced bears, a species of lion, several species of sabre-toothed cats and massive dire wolves. In Europe we had woolly mammoths and elephants, aurochs, lions, cave bears, cave hyenas, giant elks, several species of rhinoceros (including the ten-tonne indricotherium, the largest land mammal so far discovered) and much, much more. Each continent had its own magnificent selection of giant, hairy beasts.

Imagine being one of those first *Homo sapiens* to leave Africa 90,000 years ago. They were exploring a world that really did contain monsters. Every river they crossed, every new valley they entered, they could expect to stumble across new and formidable behemoths, many armed with tusks, horns, fangs and claws. As they travelled they would also have come across other hominids – tribes of creatures varying from the large and powerful, beetle-browed Neanderthals to the hobbit-like *Homo floresiensis*. Our legends are full of dragons and monsters, elves, pixies and trolls. They are not myths; we really did once live in a world that was full of such wonders.

So what happened to all these marvels? Why did they all go extinct during a relatively short period? The answer is almost certainly that we killed and ate them. Our 80,000-year journey was a culinary odyssey. Our ancestors were hunters; by working together in groups with spears and arrows, they could easily kill even the largest of the giant mammals, and a small tribe could have lived off one such kill for weeks. Their motivation for spreading to the more remote and bleak corners of our planet, such as Siberia, was probably that they were following the great herds of mammoth or bison. The animals they encountered as they spread were naïve – they had not encountered humans before and would not have known to run away. Many would have had no effective defence against missiles, so they fell easy prey. The largest creatures in particular would have bred slowly, taking many years to reach maturity (as elephants do today), and so their numbers would soon have been depleted, forcing our ancestors to move on in search of fresh hunting grounds.

Some scientists find this explanation unpalatable, and argue that periods of global cooling or epidemics of disease wiped out the megafauna, but these explanations do not fit the facts at all well, and they smack of wishful thinking. The timing of extinction in different parts of the globe closely follows our arrival, allowing a few thousand years for our populations to expand following colonisation. Humans arrived in Australia about 60,000 years ago, and discovered a land of astonishing and fabulous creatures. As well as the species found there today, there were rhinoceros-sized two-tonne wombats, at least seventeen species of short-faced kangaroo, including one that stood three metres high, and a sheep-sized echidna (spiny anteater), the largest egg-laying mammal ever to have lived. There were terrifying predators, including the leopard-sized marsupial lion, and giant predatory birds with razor-sharp, hooked bills, such as the half-tonne carnivorous 'thunder bird', and the slightly smaller but still 2.5-metre-tall *Bullockornis*, nicknamed 'the demon duck of doom' by Australian palaeontologists. Today the komodo dragon is the largest surviving species of lizard in the world, at three metres long and weighing about seventy kilos, and is a formidable beast, but it is a shrimp compared to the seven-metre-long, half-tonne monsters that the early explorers of Australia had to contend with. There was also a land-living, seven-metre-long crocodile, which is thought to have been able to run at speed and chase down its prey, and which must surely have snacked on early humans and been utterly terrifying on first (and presumably often terminal) encounter.

Despite the ferocity of some beasts, within a few thousand years of our arrival all of them were extinct. Even a seven-metre-long crocodile would have been little match for the intelligence of a group of human hunters, able to climb into trees beyond its reach, armed with sharp weapons. We made extensive use of fire to drive animals out of the thickets on to our spears, and so completely altered the landscape of Australia by denuding much of it of trees. The large animals of Tasmania had a brief reprieve as the early settlers failed to cross the wild and chilly seas of Bass Strait from southern Australia, but about 43,000 years ago a land bridge formed during a period of lower sea levels, enabling humans to cross; within 2,000 years all of Tasmania's large fauna had gone, leaving the elusive, dog-sized thylacine (also known as the Tasmanian wolf) as the largest mammal.

Extinction of the megafauna occurred later in North America, a little while after the first humans crossed over from Siberia. In the space of just a few hundred years almost all the large mammals had gone, leaving the bison as the largest survivor. Whether we deliberately hunted the large predators such as sabre-toothed tigers is not known; one might imagine that, as with the Masai of Africa, hunting and killing such fearsome creatures might have been a rite of passage for young hunters. In any case, by eliminating their prey we doomed the sabre-tooths to inevitable extinction. A few hundred years later, as humans moved southwards, the megafauna of South America was similarly extirpated.

New Zealand was colonised much more recently, about 1,000 years ago. As there were no mammals apart from bats, giant birds had evolved there, including at least eleven species of moa, the largest of which stood 3.6 metres high, the tallest bird ever to live. They must have been terribly easy to track and kill, for carbon-dating of Maori middens suggests that all eleven species were driven to extinction *within just 100 years* of man's arrival. These studies also suggest that, at least to start with, the Maoris only bothered to take the choicest cuts of meat, slaughtering the helpless beasts and leaving the bulk of their corpses where they fell.*

It is interesting that Africa is the one place left on Earth where quite a bit of the megafauna survives – elephants, giraffes, hippos, lions, and so on – given that Africa is where humans first appeared. We will never know for sure, but there is one likely explanation. Humans didn't arrive suddenly in Africa. We evolved there slowly, over millions of years, from smaller, tree-dwelling apes. As we became gradually more intelligent, more adept at making weapons, more organised in our hunting, so the wildlife of Africa had time to learn to be scared, to run away at their first sight or scent of the odd, upright apes. In contrast, the ground sloths of South America and the giant wombats of Australia would have been entirely naïve when we arrived, and were probably wiped out so swiftly that they had no time to adapt.

---

* It is sometimes argued that primitive human societies lived in harmony with nature, and that it is only modern society that is wasteful, profligate and destructive. However, the evidence suggests that humans haven't really changed much at all. Our ancestors exploited the environment as ruthlessly and with as little care for the future as we do today. The only difference is that our increased number and more advanced technology enable us to destroy the Earth much more quickly than they could manage. As Matt Ridley points out in his excellent book *The Origins of Virtue*, the idea that Native Americans had an environmental ethic and avoided over-exploiting resources was a romantic but entirely false invention of the twentieth century, later fostered by films such as *Last of the Mohicans*. Indeed, there is strong evidence that in some regions the Native Americans hunted bison simply by stampeding whole herds over the nearest cliff, only bothering to cut joints from the topmost carcasses in the pile.

It wasn't just the animals that disappeared when *Homo sapiens* spread out of Africa. So too did all other species in the genus *Homo* – our cousins. There is no clear evidence that we killed them, but it seems certain that often we did. Our record of treatment of more primitive people in recorded history is appalling (think of the Native Americans, the Aboriginal people of Australia – especially in Tasmania – or the slave trade from West Africa). It seems unlikely that our behaviour in prehistory was any better, or that our ancestors were any less violent, aggressive, belligerent and xenophobic than we are today. We probably ate many of them; in West Africa the trade in bush-meat commonly includes our closest relatives, the great apes, despite their severe endangerment, so there is no reason to suppose that we would have turned up our noses at consuming *Homo erectus* or *Homo floresiensis*. Those that we didn't kill were probably driven before us, excluded from the best hunting grounds, so that their populations dwindled. We probably didn't have it all our own way. Neanderthals were stronger than us, the top predators in Europe before we arrived, and their brains were of similar size to our own, so they would have made formidable opponents. Quite how we overcame them is not known. The Neanderthals survived in remote corners of Europe for several thousand years after our arrival, but the last of them eventually died out about 25,000 years ago. Recent controversial genetic evidence suggests that there may have been very limited interbreeding between Neanderthals and humans, and that most of us carry a few Neanderthal genes, but if so, this is all that remains of them.

Let us move on, finally and depressingly, following the trail of devastation in the wake of human diaspora, to Easter Island. This remote volcanic island, just twenty-five kilometres across, is 2,000 kilometres from its nearest inhabited neighbour, the tiny Pitcairn Island. It has a subtropical climate, and was largely forested when Polynesian settlers first arrived about 800 years ago. There were several species of tree found nowhere else in the world, including the largest known palm tree. There were also at least six species of indigenous, flightless land birds, which must have been easy to catch and good to eat; and there were nesting colonies of seabirds, providing eggs and young. The early settlers thrived, clearing land to grow crops and fishing from dugout canoes. Life was sufficiently comfortable that there was time to carve the huge statues known as Moai for which the island is famous: stylised, shadow-eyed men with jutting jaws, arranged in rows with their backs to the sea, overlooking the human settlements. The statues were carved from a single quarry and were dragged, presumably on log rollers, all over the island to each of the towns that had sprung up. The population eventually grew to about 15,000.

When the first European explorers found Easter Island in 1722, this thriving civilisation had gone.* There were no trees left anywhere on the island, the indigenous species having been eliminated to make room for crops. Without wood, the islanders had no means to make boats and so could not easily fish, or

---

* There remains some debate as to exactly what happened on Easter Island. Jared Diamond's fascinating tome, *Collapse*, paints a very bleak picture of the state of the islanders when first visited by Europeans, but his view has offended the descendants of the islanders, who resent the implication that their ancestors destroyed their ecosystem and turned to cannibalism, and argue instead that the main decline of the islanders occurred *after* European visitation. If we put political correctness to one side, the facts appear to support Diamond – there is no doubt that the islanders failed to manage the resources at their disposal, drove dozens of species to extinction and vastly reduced the capacity of the island to support life of any sort.

leave the island. They also had little from which to build houses or to burn for cooking. The indigenous flightless birds were all extinct, having long since been eaten, and the seabirds had ceased to nest, driven away presumably by over-harvesting of their eggs. Most catastrophically, without tree roots to bind the soil, much of it had blown or washed away, so that crop productivity had plummeted. As the food supply dwindled, the population seemed to have abandoned their traditional religious beliefs, perhaps feeling that their gods had deserted them. They pushed over the Moai, and turned to a new and rather violent religion known as the Bird Man Cult. By 1722 the population of 15,000 had dwindled to perhaps 2,000 or 3,000 malnourished individuals, who survived on a very limited diet of chicken, rats and – according to some – cannibalism. Their island paradise had become a bleak prison.

I am sure you can see where I am going with this. The story of Easter Island can be seen as a microcosm, a scaled-down version of what is happening in the world today. We are cutting down our forests, just as they did. Easter Island is so small that it must have been clear to all who lived there that they were rapidly using up their resources, but that did not stop them. The man who cut down the very last tree must have known that it was the last tree, must have known that without trees there would be no boats or fishing, but still he did it, presumably on the basis that his own immediate need was more important than the future of his civilisation. We too know that we are using up our resources at an unsustainable rate. We think nothing of using up fossil fuels that took millions of years to form. We know full well that we are rapidly cutting down the tropical rainforests, and that this is probably going to have a terrible impact on the Earth's climate, but we carry on doing it nonetheless. Huge areas of agricultural land around the globe have become infertile, just as they did on Easter Island. Ploughing breaks up the soil, making it easy for water to wash it away into the sea, or for wind to blow it away. Removing trees and using herbicides to kill weeds removes the roots that bind the soil together. Globally, about seventy-five billion tonnes of soil are lost every year. Forest clearance and irrigation have led to increasing salt levels in soils all over the world, in the worst cases rendering the land useless; about 320 million hectares of land have been affected by salination so far, and roughly 40 per cent of all agricultural land is now degraded in one way or another. The pesticides, fertilisers and soil particles running off into rivers destroy fresh-water aquatic communities, and when they flow out to sea they can cause tremendous harm to coral reefs and damage fish stocks, which are already under extraordinary pressure from over-fishing.

Just as the Easter Islanders drove both their native trees and flightless bird species to extinction, so on a global scale we are rapidly losing the Earth's biodiversity. Of course species have always gone extinct, long before humans came on the scene. The background extinction rate – the average rate at which species have gone extinct in the past – is estimated to be roughly one extinction per 'million species years'. This means that, if there were one million species, we would expect one to go extinct per year; or if there were only one species, we would expect it to go extinct in one million years, on average. If we estimate the current number of species on the planet to be five million, we would expect five species to go extinct each year. Of course new species also arise over time through evolution, and in the past this generally balanced or exceeded these small losses. Quantifying the current rate of extinction of the Earth's species is fraught with difficulty, not least because we do not know how many species there are on the planet. We have named one million or so, so far, but there may be anywhere between one and ten million,* with more remaining to be discovered. Proving beyond doubt that a species is extinct is also tricky; there may always be one or two hiding somewhere no one has thought to look. It is fairly easy to establish that a large creature such as a dodo, living only on one small island, has gone extinct, but for most species the task is much harder, so that only about 875 species have been officially declared extinct since the year 1500. This number is probably just a drop in the ocean compared to the actual number that has gone extinct in recent times. Current estimates, based on the rate of loss of habitat around the globe, suggest that the current extinction rate may be 100,000 times higher than the background extinction rate. Even at the most conservative end of the estimated range of extinction rates, it is likely that several species go extinct on the planet every single day. Most we have not even afforded the dignity of a name, and we will never know for certain that they ever existed. Some scientists predict that as many as two-thirds of all species on Earth will be extinct by the end of this century.

You might question whether this matters. If we never even knew they were there, who is to miss them? As the broadcaster and journalist Marcel Berlins wrote in the *Guardian* newspaper in 2008:

> Should we worry about the endangerment of all species? Pandas and tigers for sure, but armadillos? I passionately believe in saving the whale, the tiger, the orang-utan, the sea turtle and many other specifically identified species . . . Will the world and humankind be very much the poorer if we lose a thousand or so species?

---

* Some estimates of the number of species on the planet even go as high as 100 million, although an awful lot of these would be bacteria.

I hardly know where to start in explaining how misguided and ignorant this is. Berlins seems to be under the misapprehension that there is only one species of whale and sea turtle for a start, but that is a minor point. His lack of regard for the humble armadillo is disturbing – I've always found them to be rather endearing. He seems to think that species are only important if we have identified them, which presumably means that he thinks the large majority of life on Earth is irrelevant. His choice of examples suggests that the only important species are big ones, which reflects a very poor understanding of ecology, but then who said journalists need to know anything about their subject before spouting their ill-informed opinions to millions? We do not stand to lose 'a thousand or so' species, but are probably losing this many every month. The true foolishness at the heart of Berlins's statement is best explained by another quote, from Paul and Anne Ehrlich's 1981 book *Extinction*:

As you walk from the terminal toward your airliner, you notice a man on a ladder busily prying rivets out of its wing. Somewhat concerned, you saunter over to the rivet popper and ask him just what the hell he's doing. 'I work for the airline – Growthmania Intercontinental,' the man informs you, 'and the airline has discovered that it can sell these rivets for two dollars apiece.'

'But how do you know you won't fatally weaken the wing doing that?' you inquire.

'Don't worry,' he assures you. 'I'm certain the manufacturer made this plane much stronger than it needs to be, so no harm's done. Besides, I've taken lots of rivets from this wing and it hasn't fallen off yet. Growthmania Airlines needs the money; if we didn't pop the rivets, Growthmania wouldn't be able to continue expanding. And I need the commission they pay me – fifty cents a rivet!'

As the Ehrlichs go on to explain, no sane person would fly on such a plane. At some point in the future the wing will fall off, but that point might not be at all predictable. In exactly the same way that plane-rivets perform a vital role, we know that the Earth's organisms perform a whole range of important functions. Bees pollinate flowers, flies recycle dung, bacteria in root nodules fix nitrogen from the air, plants release oxygen for us to breathe, store the carbon that we release and provide us with fuel, food, clothing and drugs. Carbon and nitrogen cycles, which are vital to the health of ecosystems, involve hundreds or thousands of species, as do the processes that produce and maintain healthy soils. We rely on complex webs of interactions between species for food, clean water and clean air – interactions that we are only just beginning to understand. As with the rivets, we cannot say which species are vital and which are not. We haven't named perhaps 90 per cent of species on Earth, let alone worked out what they do. We cannot say how many species we need. What we do know is that we are losing species – popping rivets – at an unprecedented rate, and that this is reducing the ability of the Earth to support us.

There is already evidence that there are not enough pollinators to visit our crops in some parts of the world, and that as a result yields are dropping. In the apple and pear orchards of Sichuan in China, farmers have to resort to hand-pollinating every flower on the trees, sending their children clambering up to reach the flowers on the higher branches, because insects have been eradicated by the heavy use of pesticides. In India yields of insect-pollinated crops such as many vegetables are falling, due to a shortage of bees. Analyses of data from around the world by the Argentinian scientist Lucas Garibaldi have recently demonstrated that yields of insect-pollinated crops have become variable and unreliable, compared to wind-pollinated crops such as wheat. Pollination is one of the most tangible, readily explained examples of man's dependence on wildlife, but there are many more.

For all our intelligence, we do not seem to have learned from our mistakes, or seem willing to take the dire predictions of our scientists seriously. Our track record since we walked out of Africa is not good. If we continue on our current trajectory, the future is bleak, just as it was for the Easter Islanders. As we erode the capacity of the Earth to support us, so food and water shortages will become more common, probably leading to famines and wars over the dwindling resources. The human population will inevitably drop, one way or another, and that process is not likely to be a pleasant one. There will simply not be enough resources to support our large cities, and it seems likely that our civilisation will crumble. Our children will lead much poorer and harder lives than we do today.

To some extent this depressing future is unavoidable, for the damage we have already done is considerable. The Earth's climate will continue to warm for decades, regardless of whatever action we take now, leading inevitably to famine and hardship. Countless species are already extinct, or exist only in relict populations that are doomed to extinction. But that is no argument not to act – and to act now. At a global level, conservation efforts so far have been a dismal failure. We need to up our game. The sooner we stop ravaging the Earth, the less awful our future will be.

This book is intended to inspire, to encourage everyone to cherish what we have, and to illustrate what wonders we stand to lose if we do not change our ways. Biodiversity matters, in all shapes and forms. Conservation is not just about Javan rhinos and snow leopards; it is just as much about bees and beetles, flowers and flies, bats and bugs. Places such as Chez Nauche are islands where nature can thrive, but at present they are too few and far between, and they are being lost far more quickly than they are being created, particularly in the tropics, where the majority of biodiversity lives.

Go outside, look and listen. The wack-wack bird is calling. For how much longer will we hear its lonesome cry?

Epilogue

The wack-wack bird may have so far eluded identification, but the ferocious snake- and owl-eating beast finally gave up its secrets after nine years. It taunted me in the meantime . . . by leaving partly eaten corpses, not just of the owl and snake, but also of two kestrels, presumably snaffled from their roosts in the eaves of the house, and a rat. The occasional footprints in soft mud or on the Velux windows showed it to have five clawed toes, front and back.

After eight years during which we had made no progress whatsoever in identifying the beast, I decided to buy a trap. Cage traps are readily available in France – most hardware stores sell them – although I dread to think what they are generally used for and what happens to the animals they catch. My boys and I spent a summer setting the trap in places where we had seen the footprints, baited with all manner of delicacies. We tried eggs, raw meat, cooked meats and peanut butter, all to no avail. We tried peaches, grapes and apples, but the beast was not impressed. We tried cheeses – surely it could not resist my favourite, Saint Agur? It seemed that it could. Even the dormouse's favourite, Cantal cheese, drew a blank. In desperation we moved on to more exotic temptations: a selection of patisseries, including *pain au chocolat*, *tartelette au citron*, chocolate éclairs, but still with no success.

We moved the cage from place to place, trying all sorts of locations, and with an ever-growing pile of different baits in the trap. Finally, early one drizzly morning towards the end of the stay, the boys came sprinting back from an early-morning reconnoitre shouting, 'We've got it!' Success! I jumped out of bed, threw on some clothes and sprinted to where we had placed the trap, near the horse-chestnut tree at the top of the drive. Inside the trap was a large, bedraggled and very angry feral cat. I was pretty sure this was not the mythical beast – cats' claws retract, so they are not visible in footprints. Releasing the cat was a nerve-racking business as it was hissing and spitting, seemingly intent on revenge for its incarceration.

The next year I tried a different approach – a camera trap. This nifty gadget has a motion-sensitive camera and an infrared flash, so that it can photograph in darkness. We set it up, starting in one of the barns, with the camera trained on a chicken's egg. The camera has a digital display on the front, which reveals how many photographs it has taken during the night. On the first morning it had taken three pictures – although the egg was still there. It was very exciting to download the pictures on to my laptop and look through them, but there was nothing there: the camera seemed to randomly take occasional snaps, perhaps set off by passing moths or other tiny creatures.

We tried again, running through another vast selection of French titbits, to little effect. We photographed endless mice and a few voles. They at least seemed to appreciate our efforts. On one occasion a large piece of bread and peanut butter simply disappeared from right in front of the camera – there were early-evening shots, with the bait sitting there, and then shots with no bait, but no pictures of the moment when something had snuck in and made off with the bread. Clearly the camera wasn't entirely reliable, unless we were dealing with a creature that moved faster than the speed of light.

We started combining the camera trap and the cage trap – baiting the cage trap and sprinkling food all around it, and then setting up the camera to photograph anything that came near. We caught two more furious cats in the cage, and managed to photograph a nocturnal visit by a stray dog and even a roe deer, but no beast. But nine years after I purchased Chez Nauche, long after we had become resigned to the fact that the beast would elude us for ever, suddenly there it was. Four photographs, in sequence, showing a magnificent beech marten – a chunkier European relative of the pine marten – sauntering slowly around the cage trap. It was a beautiful animal, with a rich chocolate-brown coat, a creamy chest and a huge bushy tail. In the final photograph it looked disdainfully at the camera, its nose in the air, as if sniffing the wind. And then it was gone.

Acknowledgements

I must thank all of the many people that I have worked with over the last twenty years, particularly my thirty or so PhD students and the countless undergraduate project students who have working in my research group, all of whom have had to put up with my spectacularly disorganised and forgetful supervision. It was an honour to work with you all. My apologies to them and other scientists whose work I mention if there are factual errors or inaccuracies.

Particular thanks are due to my agent, Patrick Walsh of Conville & Walsh, without whom my first book, *A Sting in the Tale*, might well still be nothing more than a file on my laptop, in which case *A Buzz in the Meadow* would surely never have been written. Thanks also to my editor, Dan Franklin, and the wonderful staff at Jonathan Cape and Random House, with whom publishing is a pleasure.

Finally, I must mention Ellen Rotheray and Kirsty Park, the first people I trust to read my manuscripts. Thank you both for your encouragement, and for gently pointing out the worst of my many blunders.

Index

About the Author

DAVE GOULSON studied biology at Oxford University and is now a professor of biological sciences at the University of Stirling. He founded the Bumblebee Conservation Trust in 2006, whose groundbreaking conservation work earned him the Heritage Lottery Award for Best Environmental Project and "Social Innovator of the Year" from the Biology and Biotechnology Research Council. His previous book, *A Sting in the Tale*, was shortlisted for the Samuel Johnson Prize.